Control Systems for Beginners

Dr.D.V. Ramamurthy

Published by

BONFRING®
Intellectual Integrity

Control Systems for Beginners

ISBN 978-93-86638-78-6

Author

Dr.D.V. Ramamurthy

Bonfring

309, 2nd Floor, 5th Street Extension, Gandhipuram,

Coimbatore-641 012.

Tamilnadu, India.

E-mail: info@bonfring.org

Website: www.bonfring.org

Phone: 0422 4213231

Acknowledgement

The author expresses his deep sense of gratitude to teachers for their blessings and teachings. In particular, thanks to Dr.B.C. Nakra for the course Automatic Controls at IIT Delhi through which I began my journey in the area of control systems. Also, I am ever indebted to Dr. Prabhakar R. Pagilla for the mentorship and his patience with me. All that is good in these pages is but a reflection from my teachers and all the blemishes are purely from me.

Support from all my family members and blessings of my parents are the sources of energy which are taking me to places which I could not have dreamt of through my own mental capabilities alone.

Dr.D.V. Ramamurthy

Author Profile

Dwivedula Venkata Ramamurthy, graduated from Andhra University College of Engineering, Visakhapatnam with a B.E. degree in Mechanical Engineering in the year 1987, from Indian Institute of Technology, Delhi with M.Tech. in Design of Mechanical Equipment in the year 1992, and from Oklahoma State University, Stillwater, OK, USA with Ph.D. degree in Mechanical Engineering, in the year 2005. Ramamurthy worked extensively in the area of Control Systems Theory and implementation of control algorithms. Ramamurthy has to his credit, over 23 years of experience in teaching Control Systems and Theory of Machines.

References

1. Katsuhiko, O, Modern Control Engineering, Pearson Publishers, 5th Edition, 2015.

2. Farid, G and Benjamin C.K, Automatic Control Systems, Wiley Publishers, 9th Edition, 2014.

3. Nagrath, J and Madan, G, Control System Engineering, New Age International Publishers, 5th Edition, 2009.

<table>
<tr><th>Unit</th><th>Contents</th><th>Page No</th></tr>
</table>

ప్రార్ధన

ప్రభాకర నమస్తుభ్యం దివాకర నమోస్తుతే

జ్ఞాన ప్రదాతాయ నమస్తుభ్యం ప్రసీద మమ భాస్కర

ఒగాటాయ నమః

కువో దేవాయ నమః

డార్వాయె నమః బిషఫె నమః

నగరతాయ నమః

మదన గోపాలాయ నమః

విద్యా సాగరాయ నమః

మిసావాయ నమః

గేరీయంగాయ నమః

గేరీ యెన్ దేవాయ నమః

మార్టిన్ హెగన్ దేవాయ నమః

CONTROL SYSTEMS FOR BEGINNERS

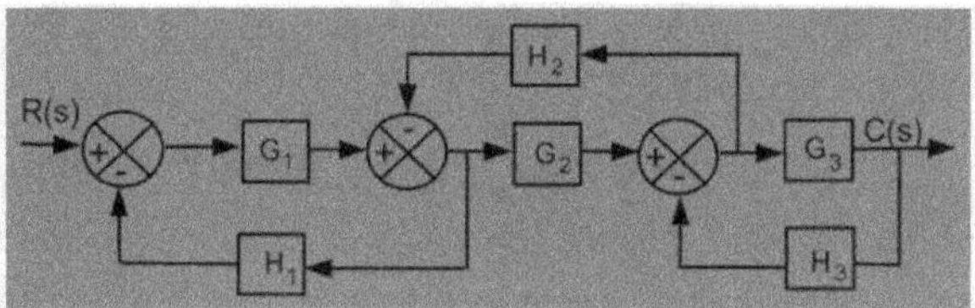

1. Why This Book

There are innumerable books on Control Systems and many more learning resources including video lectures. One may then wonder: "Why another book?" Well, as they say, variety is the spice of life. And I fervently hope that this book has to offer at least slightly different variety of presentation. There is yet another reason for this book. Modern-day student sometimes needs a quick refresher to go through what he has learnt all through the semester, perhaps in a concise book, full of pictures and gentle explanation for him to cull through quickly for his exam preparation. This book is expected to serve that purpose. Also this book is expected to help the modern-day teacher who wants to glance through the side-headings and brief background before stepping into the class. All the topics are presented in almost colloquial language and tone so that the casual reader can glance through the book without any intimidation. Yet the topics presented do not lack rigor. The book has several pictures and explanations so that visual learning is possible. The book contains topics of classical control theory present in the curriculum of many universities and omits state-space methods. Perhaps in time to come, those topics will be included in another version of the book.

In passing, let me mention that this book is intended to be a "starter" before going into the final dinner of standard text books. It is my advice to all the students to purchase at least one standard text book on Control Systems (such as the ones authored by Ogata, Kuo, Dorf & Bishop, or Nagrath & Gopal) and read through the text book patiently. If you like this book, please let me know. If you do not like this book, then let me know too! My email address is ramdwivedula@gmail.com.

1.1. Control Systems: Introduction

Engineering is concerned with understanding and controlling the materials and forces of nature for the benefit of humankind. And the subject "Control Systems" deals with systems with a view to obtain desired performance out of "Systems" by designing them according to the specifications and/or modifying their performance by inserting appropriate "Control" elements into the system.

In this context, we use the word "System" to mean collection of components and/or subsystems so formed, arranged, and connected as to serve a useful purpose. Examples of systems include: A table fan, a refrigerator, light bulb, power plant, microwave oven, etc.

Thus, "Control System" is an interconnection of components arranged in such a way that desired response is obtained from the system we are dealing with.

Fortunate control system engineers have the opportunity to design control systems and develop control algorithms to control machines, industrial processes, economic processes for the benefits of the society.

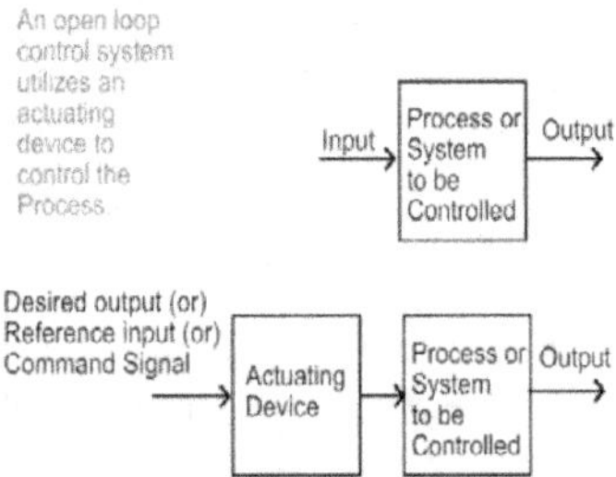

When we compute the desired input, give that input to the system, and derive output from the system without verifying whether the output is exactly the desired one or not, we say that the system is operating in "open loop mode." Many household appliances operate in open-loop mode. For example the fan, the bread-toaster, and gas stove are typically used in open-loop mode. Figure below shows system in open loop mode.

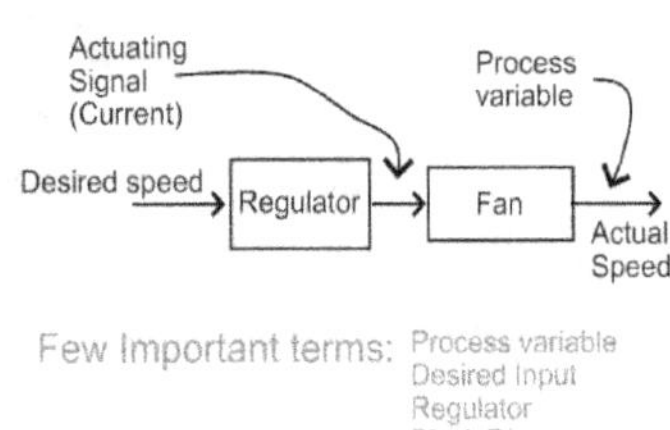

Few Important terms: Process variable
Desired Input
Regulator
Block Diagram

Example 1: Open Loop System, Fan

Let us say our "system" is fan and the process variable of interest to us is speed of fan. At different times of the we are interested in setting the speed of the fan to different values of speed as desired by us. To achieve this we use a component called "<u>regulator</u>" with markings 1, 2, 3, 4, 5, etc. Let us say that marking 1 corresponds to 50 rpm, 2 correspond to 100 rpm and so on. This idea can be shown in a figure as shown below.

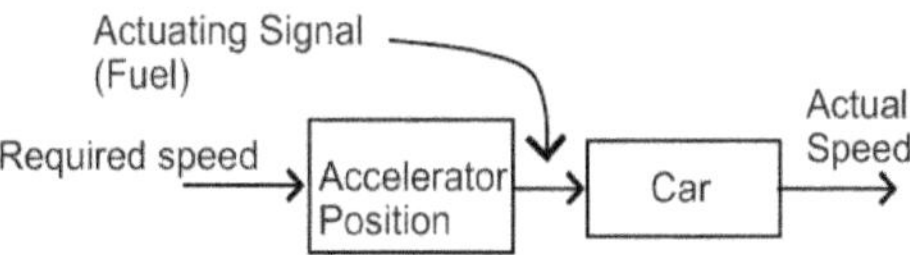

Example 2: Open Loop System, Car

Let us consider our system to be a car and the process variable to ne speed. In a simplified way, omitting many details we may represent the situation in a block diagram as shown below.

Few important terms: Process variable, desired input, Actuating signal, Regulator, block diagram.

The situation is fine so far and in many situations this may be more than satisfactory. But real-life is not as simple as it appears! We will always have disturbances that come in the way of achieving our desired process variable.

For example, the car in Example-2 is considered here. On flat, level road, a given accelerator position gives a given speed. But there are several disturbances such as wind speed, friction at road, slope of road etc which are not directly in our control. These disturbances necessitate intervention. For example the <u>driver</u> may have to give more acceleration or less. Since these disturbances are unavoidable and often unpredictable, the driver has to be very cautious if he has to maintain the speed constant. Also note that driver has to keep on looking at the speedometer to ascertain that he (or she) is going at desired speed: the driver has to keep moving his foot to press or release the accelerator pedal, other than handling the steering wheel and clutch. This is ok for those who enjoy it. But what if the driver is bored and tired of pressing the accelerator pedal? What if a system takes care of maintaining the speed while the driver enjoys the drive steering the car and listening to music? This thought gave rise to "Automatic Control System" or "Feedback Control System".

Example 3: Closed Loop System

The situation shown above gives us line of thought and sequence of events that occur in the case of controlling the speed of car. Let us say that the driver wants to go at 90 kmph[1]. This is <u>desired speed</u> and the driver must think of this speed always – throughout the duration of the

---

[1] This is much above the speed limit on Indian highways. So don't try to drive at this speed.

driving and <u>compare</u> the speed of the car with 90 kmph. If the actual speed of the car shown on speedometer is different, there is <u>error</u>. Depending on error, the driver issues <u>control signal</u> by pressing or releasing accelerator.

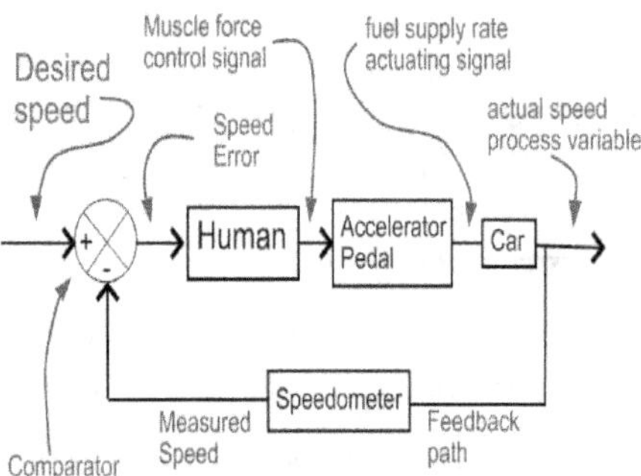

When accelerator position changes, more (or less) fuel goes in and car either accelerates or decelerates. Hence speed changes. The driver sees the <u>speedometer</u> and <u>compares</u> it with what 90 kmph. The cycle goes on and on and on. Going through the sequence of events again and again <u>continuously</u> may (and will) become tiresome to the driver.

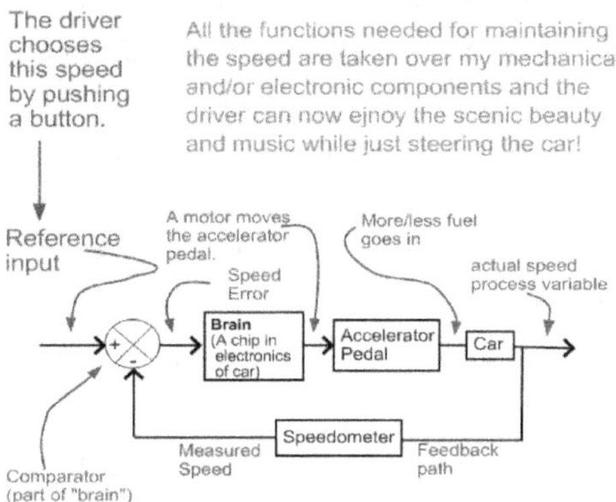

As a control system engineer can we do something to alleviate this burden? The answer to this question is, fortunately, yes, and we have <u>cruise control systems</u> in modern cars. A block diagram of cruise control system may be drawn as shown beside. The "thought process" required for maintaining the speed is taken over by chip called "the brain" of the car. The driver is relieved of the burden. This is one function of "brain" and as cars get more and more sophisticated, the "brain" in the car takes more and more functions. One day we may have driver-less cars !

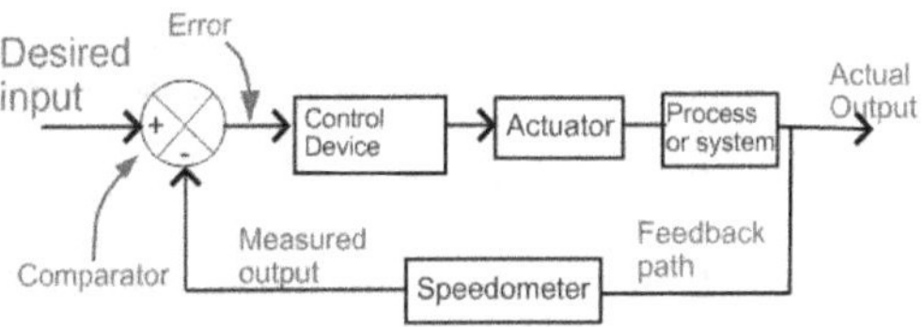

As seen in the block diagram above, every function performed by human is replaced by a Mechanical (or electronic) component. When this is done, we have an <u>Automatic Control System</u>. In fact conceptually, every Automatic Control System or <u>Feedback Control System</u> can be shown in <u>a block diagram</u> as given below. Such systems are called <u>closed loop systems</u>.

It is important to be able to draw such conceptual block diagrams of control systems because these will be useful while drawing <u>mathematical block diagrams.</u>

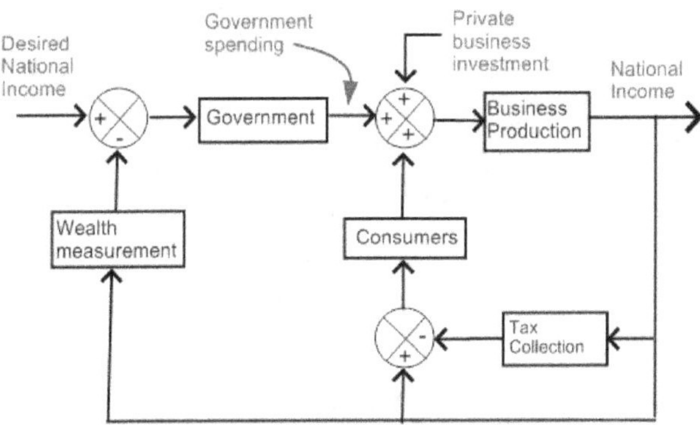

Home Work 1

(Q1).   Differentiate between "regulation" and "Control" in the context of course.

(Q2).   Figure below shows National Income Feedback Control System. Carefully explain each block and sequence of events related to this block diagram.

Can you identify the recent reforms proposed by Indian Government and explain where they fit into the block diagram above?

(Q3).   A particular household, Mrs. Dutta's family, has a bore-well from which they draw water and fill an overhead water tank. Very often when they forget to switch off the motor of the pump, water overflows causing wastage and inconvenience. Can you, as a Control Engineer, provide a solution? Draw a Conceptual block diagram of system you propose.

(Q4).   Think of any two situations that need accurate control of process variables and explain the problems involved. Draw relevant Conceptual Block diagrams.

1.2.   Classification of Control Systems

Control systems can be classified in several ways and what is given hereunder is an overall view.

Linear and Non-Linear Systems

A system is said to be linear if principle of super position applies to it for inputs with zero initial conditions.

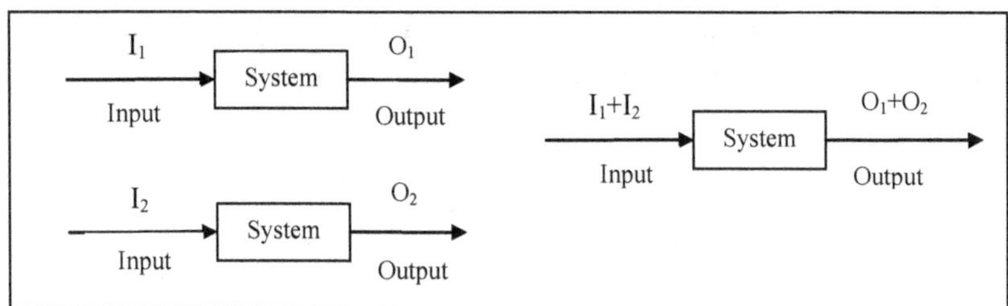

Most, if not all, systems are non-linear in nature because one or more of non-linearities such as saturation, friction, dead zone and backlash exist. However, for limited range of operation of system, it is possible to linearize the underlying non-linear equations and arrive at a linearized system.

Continuous Time and Discrete Time Control System

In continuous time control systems, all systems variables are the functions of a continuous time variable "t".

In discrete time systems, one or more system variables are known (or measured or acquired) only at certain discrete intervals of time.

Typically the underlying dynamic equations describing a continuous time system are differentiated differential equations and underlying equations of discrete time system are difference equations.

Continuous time systems are also often referred to as analog systems.

Discrete time systems are also referred as digital systems, computer controlled systems, or sampled data systems.

Deterministic and Stochastic Control Systems

A control system is said to be deterministic when its response to input as well as behaviour to external disturbances is <u>predictable</u> and <u>repeatable</u>.

If such response is unpredictable, the system is said to be stochastic in nature.

Most of the systems in nature are stochastic in nature. However, if we are not <u>fussy</u> about exactness of measured values, many systems may be considered as deterministic. stochastic systems re described by <u>axioms</u> of probability and <u>random variables</u>.

Lumped Parameter and Distributed Parameter Systems

In a lumped parameter systems, the system parameters are considered to be located at determined points and equations are developed to describe the interrelationships.

In a distributed parameter system, the system parameters are considered to be <u>spread</u> across the system <u>spatially</u>.

In real life, most of the systems are distributed systems and are slightly difficult to handle mathematically because they need to be described by <u>partial differential equations</u>. But, with reasonable accuracy, these systems can be <u>modelled</u> as lumped parameter systems and can be represented by ordinary differential equations.

SISO and MIMO Systems

A system that has only one input and one output (<u>S</u>ingle <u>I</u>nput and <u>S</u>ingle <u>O</u>utput) is called a SISO system; otherwise it is called <u>M</u>ultiple <u>I</u>nput <u>M</u>ultiple <u>O</u>utput system.

Open Loop and Closed Loop System

A system in which is output is dependent on input <u>controlling action</u> (or input) is totally independent of output or changes in output is called as open loop system.

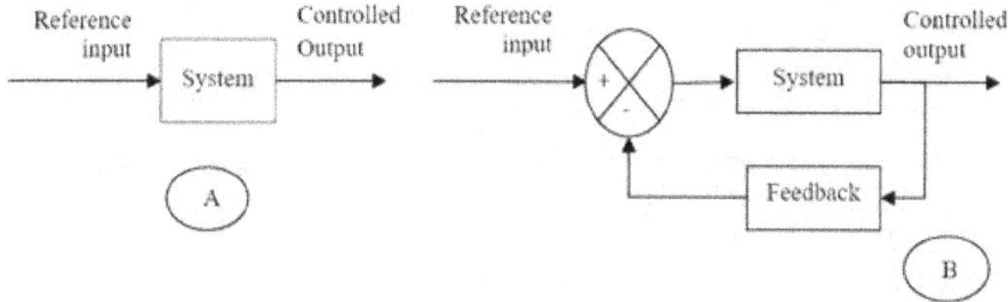

A system in which control input is somehow dependent on the output or changes in the output is called closed loop system.

In this block diagram descriptions, if we follow the directions of the arrows, we see that, in system B, the <u>control path</u> appears "closed". Hence it is called <u>Feedback system</u> or <u>closed system</u>. Whereas there is no "closed loop" in system A, hence system A is called "open loop" system.

Time-Varying and Time-Invariant System

Time-varying control systems are those in which parameters of the system are varying with time. If parameters are constants, the system is called Time-invariant.

Time-invariant systems are described by either differential or difference equations with constant coefficient. Whereas time-varying systems, the underlying equations have coefficients which are not constants. In this course, we are concerned with Linear, Time-invariant, lumped-parameter, continuous-time, deterministic, SISO systems. When everything else is defined or understood we are simple call them LTI system (<u>L</u>inear <u>T</u>ime <u>I</u>nvariant).

The type of systems we deal with in this course are described by (or approximately modelled using) linear differential equations with constant coefficients such as:

$$a_n \frac{d^n y}{dt^n} + a_{n-1} \frac{d^{n-1} y}{dt^{n-1}} + \cdots + a_1 \frac{dy}{dt} + a_0 y_1$$

$$= b_m \frac{d^m u}{dt^m} + b_{m-1} \frac{d^{m-1} u}{dt^{m-1}} + \cdots + b_1 \frac{du}{dt} + b_0 u$$

In this equation, the coefficients a_n, a_{n-1},......, a_1, a_0 and b_m,......, b_0 are constants and depend only on system parameters, y is the output from the system, and u is the input to the system.

If it is required that the output at any time instant depend on input or prior inputs only, then, an essential requirement is that m ≤ n . This happens in real-life system, and such systems are called as <u>causal systems</u>. In our course, we deal with only casual systems.

However, do NOT be led to the conclusion that non-casual systems do Not have a purpose. In image processing applications, we may use differential equations of type (A) with m ≥ n. Such applications may be termed as <u>non-causal filters</u>.

The rest of the course deals extensively with systems that are modeled by equations of type (A), solutions of which are very convenient using Laplace Transforms. So, we review Laplace Transforms briefly in next couple of classes.

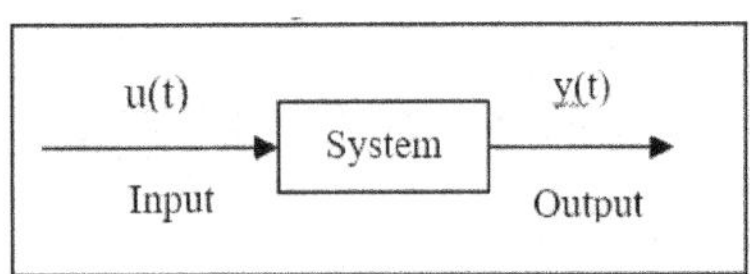

Home Work 2

(Q1). A system shown in block diagram below is defined by y(t) = cos[u(t)]. Is this system time invariant?

(Q2). Is the system shown in (Q1) above, a linear system?

(Q3). A continuous time system is described by mathematical equation y(t) = u(t+1) Where u(t) is input to the system and output from the system is y(t) . Is this a casual system?

(Q4). A system has input of u(t) and its output at anytime instant is given as: y(t) = 4u(t)+3 . Can this system be called as linear system?

(Q5). The differential equation describing a system is given by

$$t\frac{dy\,(t)}{dt} + 7y(t) = 2u(t)$$

Where u(t) is input at time instant t and y(t) is output at time instant t.

1.3. Laplace Transform

For the type of systems and situations we are interested in analyzing and exploring, the underlying equations describing the dynamics are of the form.

$$a_n \frac{d^n y}{dt^n} + a_{n-1} \frac{d^{n-1} y}{dt^{n-1}} + \;-\;-\;-\;-\; + a_1 \frac{dy}{dt} + a_0\, y$$

$$= b_m \frac{d^m u}{dt^m} + b_{m-1} \frac{d^{m-1} u}{dt^{-1}} + \;-\;-\;-\;-\; + b_1 \frac{du}{dt} + b_0\, u \;\text{------- (A)}$$

Where $a_n, a_{n-1}, \ldots, a_1, a_0$ and $b_n, b_{n-1}, \ldots, b_1, b_0$ are real numbers which are system parameters in some sense, y is the output of the system and u is the input of the system. Both u(t) and y(t) are assumed to be continuous functions of time(t). Further m≤n for <u>causality</u>.

To be able to predict the behavior of the system in different scenarios, we must be able to solve the differential equation given above with a set of initial conditions. This problem is referred to as <u>Initial Value Problem</u> (IVP).

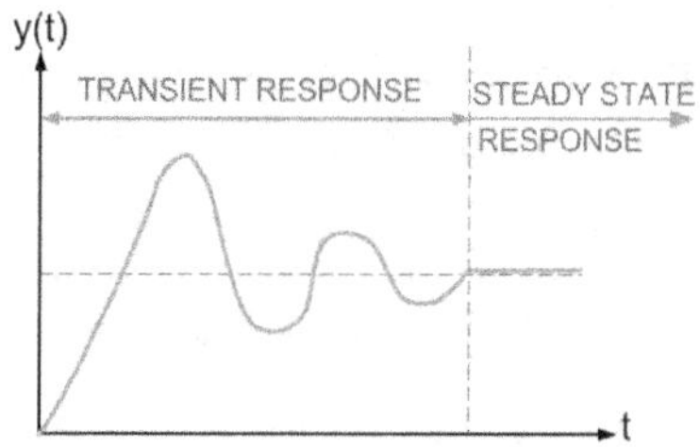

We know that the Solution of (A) for a given u(t) consists of two parts: Complimentary Function (CF) and Particular Integral (PI). The CF indicates <u>Transient Response</u> of the System, within a short span of time and PI indicates <u>Steady State Response</u>, after a relatively long period of time. Figure beside shows a <u>step-response</u> of a system. Typically the response involves an oscillatory initial part and relatively non-oscillatory "settling part".   Clear specifications of the oscillatory part and the settled part form transient and steady-state response of the system. As discussed previously in this paragraph, mathematically, these are derived from CF and PI.

The usual "Calculus Method" of finding solution involves finding CF and PI and write y(t)=CF+PI. This method, though straight forward may be a   cumbersome for even slightly larger values of m and n.

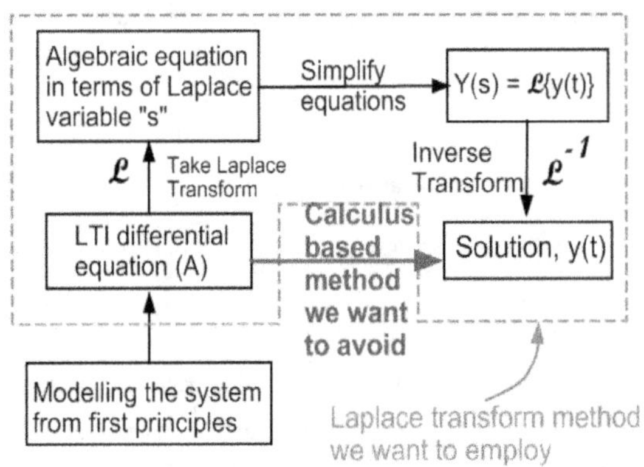

To <u>circumvent</u> this problem, we resort to Laplace <u>Transforms</u> as shown in figure beside.

To employ the Laplace Transform method shown in the pink, dotted lines in diagram and <u>obviate</u> the calculus based method shown in blue, bold we must first understand a few points about Laplace Transforms.

We will state important definitions, Theorems, and results concerning Laplace Transforms and provide a few examples of Laplace Transforms.

Definition

Laplace Transform of a function f(t) exists or f(t) is <u>transformable</u> if

$$\int_{0-}^{\infty} |f(t)| e^{-\sigma_1 t} dt < \infty$$

For some real, positive σ_1. It may be noted that signals that are physically realizable always have Laplace Transform. The Laplace Transform, F(s), for a function of time f(t) is defined as:

$$F(s) \triangleq \int_{0^-}^{\infty} f(t)\, e^{-st}\, dt = \pounds\,[f(t)]$$

Let G(s) be the Laplace transform of g(t) so that $\pounds[g(t)] = G(s)$; a, c_1, c_2, $\alpha \in R$ be constants. Then, we can list properties of the Laplace operator $\pounds$ as:

Properties of Laplace Operator

(i)   **Linearity**: $\pounds[c_1 f(t) + c_2 g(t)] = c_1 F(s) + c_2 G(s)$

(ii)   **Time Delay**: $\pounds[f(t-a)\, u_s(t-a)] = e^{-as} F(s)$

Where unit step function $u_s(t)$ is defined as:

$$u_s(t) = \begin{cases} 0 \; if \; t < 0 \\ 1 \; if \; t \geq 0 \end{cases}$$

(iii)   **Derivative Property**: $\pounds\left[\dfrac{df}{dt}\right] = sF(s) - f(0)$

$$\pounds\left[\frac{d^2 f}{dt^2}\right] = s^2 F(s) - sf(0) - f^1(0)$$

$$\pounds\left[\frac{d^n f}{dt^n}\right] = s^n F(s) - \sum_{i=1}^{n} s^{n-i} f^{(i-1)}(0)$$

The notation used is $f^i(0) = \dfrac{d^i f}{dt^i}\Big|_{t=0}$

(iv)   **Integration**: $\pounds\left[\int_0^t f(\lambda)d\lambda\right] = \dfrac{F(s)}{s}$

(v)   **Convolution**: $\pounds[f(t) * g(t)] = F(s)G(s)$ where $*$ is convolution operator defined by

$$f(t) * g(t) = \int_{-\infty}^{\infty} f(\lambda) g(t - \lambda) d\lambda$$

(vi)   **Initial Value Theorem**: $\mathrm{Lim}_{t \to 0}\, f(t) = \mathrm{Lim}_{s \to \infty}[sF(s)]$

(vii)   **Final Value Theorem**: $\mathrm{Lim}_{t \to \infty}\, f(t) = \mathrm{Lim}_{s \to 0}[sF(s)]$

(viii)   **Multiplication by time**: $\pounds[t\, f(t)] = -\dfrac{dF}{ds}$

(ix)   **Complex shift** : $\pounds[e^{-\alpha t} f(t)] = F(s + \alpha)$

(x)   **Time Scaling**: $\pounds\left[f\left(\dfrac{t}{\alpha}\right)\right] = \alpha\, F(\alpha s)$

Students are encouraged to commit these TEN properties to memory and practice at least two examples for each property.

Example-1:- $\pounds[1] = \int_0^{\infty} e^{-st}.1\; ds = \tau\left[\dfrac{-e^{-st}}{s}\right]_0^{\infty} = \dfrac{1}{s}$ using definition of Laplace Transform

Example-2:- Let us use property (viii) to find Laplace Transform of t. Let $f(t) = 1$ then $F(s) = \frac{1}{s}$ as shown in example (1).

$$\text{Then } £[t\, f(t)] = -\frac{dF}{ds} = \frac{1}{s^2}$$

Example-3:- Let us use property (ix) to find Laplace Transform of $e^{-\alpha t}$

$$\text{Let } f(t) = 1 \text{ then } F(s) = \frac{1}{s}$$

$$\text{Then } £\left[e^{-\alpha t} f(t)\right] = £[e^{-\alpha t}] = F(s + \alpha) = \frac{1}{s+\alpha}$$

Example-4:- Let us find Laplace Transform of $3 + 2t + \frac{e^{-at}}{5}$.

$$\text{Let } f(t) = 1,\ g(t) = t \text{ and } h(t) = e^{-at}$$

$$\text{Then } £\left[3 + 2t + \frac{e^{-at}}{5}\right] = £\left[3f(t) + 2g(t) + \frac{1}{5}h(t)\right]$$

$$= 3F(s) + 2G(s) + \frac{1}{5}H(s) \text{ using Linearity (i)}$$

$$= \frac{3}{s} + \frac{2}{s^2} + \frac{1}{5(s+\alpha)}$$

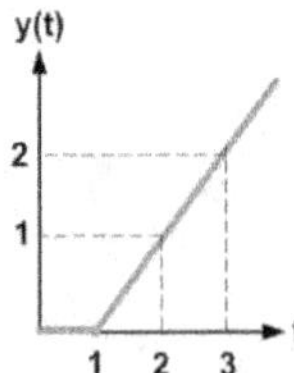

Example-5:- Find the Laplace Transform of the function g(t) shown in graph

The g(t)-plot shown in graph may be written as:

$$g(t) = f(t - 1)u_s(t - 1) \text{ where } f(t) = t.$$

We know that example (2), $F(s) = £[f(t)] = \frac{1}{s^2}$

$$\therefore £\,[g(t)] = £[f(t - 1)u_s(t - 1)] = \frac{e^{-s}}{s^2}$$

Home Work 3

(Q1). If $f(t) = \begin{cases} 2 & \text{of } t \in [0,2] \\ 0 & \text{other wise} \end{cases}$ find $£[f(t)]$. $\left[\text{Answer:}\frac{2}{s}[1 - e^{-2s}]\right]$

(Q2). Find Laplace Transform of $f(t) = t^2 e^{-2t}\cos 3(t)$

$$\left[\text{Answer:}\frac{2(s + 2)(s^2 + 4s - 23)}{(s^2 + 4s + 13)^3}\right]$$

(Q3). Find the Laplace Transform of $\cos^2(3t)$. $\left[\text{Answer:}\frac{1}{2s} + \frac{s}{2(s^2+36)}\right]$

(Q4).   Find the Laplace Transform of $f(t) = (e^{-2t} - 1)^2$

$$\left[Answer = \frac{8}{s(s+2)(s+4)}\right]$$

(Q5).   Find the Laplace Transform of f(t)=sin(2t) cos(2t)

$$\left[Answer = \frac{2}{(s^2 + 16)}\right]$$

Inverse Laplace Transform

1. *Definition*

When $F(s) = \int_0^\infty e^{-st} f(t)dt$,

We may say that $f(t) = £^{-1}[F(s)]$.

We may also write $f(t) = \frac{1}{2\pi j}\int_{\sigma-j\infty}^{\sigma+j\alpha}$

We will see the properties of inverse Laplace Transform and a few problems in next class.

2. *Inverse Transform Properties*

1) **LINEARITY**:- If c_1 and c_2 are constants while $F_1(s)$ and $F_2(s)$ are Laplace Transforms of $f_1(t)$ and $f_2(t)$, then,

$$£^{-1}\{c_1F_1(s) + c_2 F_2(s)\}= c_1f_1(t) + c_2f_2(t)$$

2) **First Translation**:- If $£^{-1}\{F(s)\}= f(t)$ then $£^{-1}\{F(s-a)\}= e^{at} f(t)$

3) **Second Translation**:- If $£^{-1}\{F(s)\}= f(t)$ then $£^{-1}\{e^{-as} F(s)\}= g(t)$

Where

$$g(t) = \begin{cases} f(t-a) \ when \ t > a \\ 0 \qquad when \qquad t \le a \end{cases}$$

We may write g(t)= f(t-a)u_s(t-a)   where $u_s(t)$ is <u>unit step input.</u>

4) **Change of scale**:- If $£^{-1}\{F(s)\}= f(t)$ then $£^{-1}\{F(as)\}= \frac{1}{a}f\left(\frac{t}{a}\right)$

5) **Derivatives of Transform**:- If $£^{-1}[F(s)]=f(t)$ then $£^{-1}\left\{\frac{d^n}{ds^n}F(s)\right\}= (-1)^n t^n f(t)$

6) **Integral of Transform** :- If $£^{-1}[F(s)]= f(t)$ then $£^{-1}\int_s^\infty F(u)du] = \frac{f(t)}{t}$

7) **Multiplication by s**:- If $£^{-1}[F(s)]= f(t)$ then

$$£^{-1}\{sF(s)\} = \frac{df}{dt} \ if \ f(0) = 0$$

8) **Divison by s^n** :- If $£^{-1}\{F(s)\}= f(t)$ then $£^{-1}\left[\frac{F(s)}{s}\right] = \int_0^t f(u)du$

Example 1

$$- £^{-1}\left[\frac{3(s^2-2)^2}{2s^5}\right] = £^{-1}\left[\frac{3(s^4-4s^2+4)}{2s^5}\right]$$

$$= \frac{3}{2}\pounds^{-1}\left\{\frac{1}{s}\right\} - 6\pounds^{-1}\left\{\frac{1}{s^3}\right\} + 6\pounds^{-1}\left\{\frac{1}{s^5}\right\}$$

$$= \frac{3}{2} - 3t^2 + \frac{1}{4}t^4$$

Example 2

$$:- \pounds^{-1}\left\{\frac{4s+12}{s^2+8s+16}\right\} = 4\pounds^{-1}\left\{\frac{1}{s+4}\right\} - 4\pounds^{-1}\left\{\frac{1}{(s+4)^2}\right\}$$

$$= 4e^{-4t} - 4te^{-4t}$$

$$= 4e^{-4t}(1-t)$$

Home Work 4

(Q1). Find $\pounds^{-1}\left\{\frac{s+1}{s^2-6s+25}\right\}$ Answer: $e^{-3t}\left[\cos(4t) - \frac{1}{2}\sin(4t)\right]$

(Q2). Find $\pounds^{-1}\left\{\frac{3s+1}{(s+1)^2}\right\}$ Answer: $e^{-t}\left[\frac{3t^2}{2} - \frac{t^3}{3}\right]$

(Q3). Find $\pounds^{-1}\left\{\frac{6}{2s-3} - \frac{3+4s}{96s^2-16} + \frac{8-6s}{16s^2+9}\right\}$

         Answer: $3e^{3t/2} - \frac{1}{4}\sinh\left(\frac{4t}{3}\right) - \frac{4}{9}\cosh\left(\frac{4t}{3}\right) + \frac{2}{3}\sin\left(\frac{3t}{3}\right)$

(Q4). Find $\pounds^{-1}\left\{\frac{1}{s^2+6s+13}\right\}$ Answer: $\frac{1}{2}e^{-3t}\sin(2t) - \frac{3}{8}\cos\left(\frac{3t}{4}\right)$

(Q5). Find $\pounds^{-1}\left\{\frac{10}{(s-5)^2} + \frac{2}{s-5^3}\right\}$ Answer: $10te^{st} + t^2e^{5t}$

1.4. Solving Linear Ode using Laplace Transform

The general method for solving a linear ODE involves:

1. Taking Laplace Transforms
2. Algebraic simplification for obtaining Laplace Transform of dependent variable
3. Taking inverse Transform

Example 1

Solve $\frac{d^2y}{dt^2} - 5\frac{dy}{dt} + 6y = 0$ with initial conditions $y = 2$, $\frac{dy}{dt} = 2$ *when* $t = 0$

Solution

Take Laplace transforms on both sides

$\pounds\left[\frac{d^2y}{dt^2}\right] - 5\pounds\left[\frac{dy}{dt}\right] + 6\pounds[y] = 0$ *Let* $\pounds[y] = Y(s)$ then, we know from the properties of Laplace transforms:

$$\pounds\left[\frac{d^2y}{dt^2}\right] = s^2Y(s) - sy(0) - y'(0) = s^2Y(s) - 2s - 2 \text{ and}$$

$$\pounds\left[\frac{dy}{dt}\right] = sY(s) - y(0) = sY(s) - 2$$

$$\therefore s^2Y(s) - 2s - 2 - 5[sY(s) - 2] + 6Y(s) = 0$$

$$\Rightarrow (s^2 - 5s + 6) + Y(s) = 2s - 8$$

$$\Rightarrow Y(s) = \frac{2s - 8}{s^2 - 5s + 6} = \frac{A}{s - 2} + \frac{B}{S - 3}$$

$$\Rightarrow A(s - 3) + b(s - 2) = 2s - 8$$

$put\ s = 3\ to\ get\ B = -2$

$put\ s = 2\ to\ get, A = 4$

$$\Rightarrow Y(s) = \frac{4}{s - 2} - \frac{2}{S - 3}$$

Take inverse transform to get $y(t) = 4e^{2t} - 2e^{3t}$

Example 2

$:-\dfrac{d^2y}{dt^2} + \dfrac{dy}{dt} + 5y = 10e^t$ with initial conditions $y = 1$, $\dfrac{dy}{dt} = 2\ at\ t = 0$

Solution

$$£\left[\frac{d^2y}{dt^2}\right] - 4£\left[\frac{dy}{dt}\right] + 5£[y] = £[10e^t] = \frac{10}{s - 1}$$

$$\Rightarrow \left\{s^2 Y(s) - sy(0) - y'(0)\right\} + 4\{sY(s) - y(0)\} + 5Y(s) = \frac{10}{s - 1}$$

$$\Rightarrow (s^2 + 4s + 5)Y(s) = \frac{10}{s - 1} + s + 6$$

$$\Rightarrow Y(s) = \frac{s+6}{(s^2+4s+5)} + \frac{10}{(s-1)(s^2+4s+5)}$$

$$Let\ \frac{10}{(s - 1)(s^2 + 4s + 5)} = \frac{A}{s - 1} + \frac{Bs + c}{(s^2 + 4s + 5)}$$

$$\Rightarrow A(s^2 + 4s + 5) + (s - 1)(Bs + c) = 10$$

Equate coefficients of s^2: $A + B = 0$ $\hspace{4cm}$ (1)

Equate coefficients of s: $4A + C - B = 0$ $\hspace{3cm}$ (2)

Equate constant terms: $5A\text{-}C=10$ $\hspace{4cm}$ (3)

$\hspace{1cm}$ $(1) \Rightarrow A = -B.\ So$ using $(2), C = 5B,$ So using $(3)\ B = -1.\ So\ A = -B = 1.$

$$\therefore Y_1(s) = \frac{s + 6}{s^2 + 4s + 5} + \frac{1}{s - 1} - \frac{s + 5}{s^2 + 4s + 5} = \frac{1}{s - 1} + \frac{1}{(s + 2)^2 + 1}$$

Taking inverse transforms, $y_1(t) = e^t + e^{-2t}\sin(t)$.

Similarly, we see that $Y_2(s) = \dfrac{s+6}{(s^2+4s+5)} = \dfrac{s+2}{(s+2)^2+1^2} + 2\dfrac{2}{(s+2)^2+1^2}.$

Therefore, $y_2(t) = e^{-2t}(\cos t + 2\sin t)$. Thence, we see that,

$$y(t) = y_1(t) + y_2(t) = e^t + e^{-2t}(\cos t + 3\sin t).$$

Home Work 5

(Q1). Solve the initial value problem: $\frac{dy}{dt} = e^{-3t}$ given that y(0)= 4 $\left[Answer: \frac{13}{3} - \frac{1}{3}e^{-3t}\right]$

(Q2). Solve the initial value problem : $\frac{dy}{dt} + y = e^t$ given that y(0)= 1 $\left[Answer: \frac{1}{2}[e^t + e^{-t}]\right]$

(Q3). Solve the initial value problem : $\frac{d^2y}{dt^2} + 2\frac{dy}{dt} + 2y(t) = 0$ with initial conditions y(0) = +1; $\dot{y}(0)$ = -1

(Q4). Solve the differential equation $\frac{d^2y}{dt^2} + y(t) = \sin(t)$ with the conditions y(0) = 0, $\dot{y}(0)$= 3 $\left[Answer: \frac{-1}{2}tcost + \frac{1}{2}sint + 3sint\right]$

Table 1: Laplace Transform Pairs

f(t)	F(s)
Step function, u(t)	$\dfrac{1}{s}$
e^{-at}	$\dfrac{1}{s+a}$
$\cos \omega t$	$\dfrac{s}{s^2 + \omega^2}$
t^n	$\dfrac{n!}{s^{n+1}}$
$f^{(k)}(t) = \dfrac{d^k f(t)}{dt^k}$	$s^k F(s) - s^{k-1}f(0^-) - s^{k-2}f^1(0^-) - \cdots - f^{(k-1)}(0^-)$
$\displaystyle\int_{-\infty}^{t} f(t)dt$	$\dfrac{F(s)}{s} + \dfrac{1}{s}\int_{-\infty}^{0} f(t)\,dt$
Impulse function $\delta(t)$	1
$e^{-at}\sin\omega t$	$\dfrac{\omega}{(s+a)^2 + \omega^2}$
$e^{-at}\cos\omega t$	$\dfrac{(s+a)}{(s+a)^2 + \omega^2}$
$\dfrac{1}{\omega}[(\alpha - a)^2 + \omega^2]^{1/2}e^{-at}\sin(\omega t + \emptyset), \emptyset = tan^{-1}\dfrac{\omega}{\alpha - a}$	$\dfrac{(s+\alpha)}{(s+a)^2 + \omega^2}$
$\dfrac{\omega}{\sqrt{1-\zeta^2}}e^{-\zeta w_n t}\sin\omega_n\sqrt{1-\zeta^2}\,t, \zeta < 1$	$\dfrac{\omega_n^2}{s^2 + 2\zeta\omega_n s + \omega_n^2}$
$\dfrac{1}{a^2 + \omega^2} + \dfrac{1}{\omega\sqrt{a^2 + \omega^2}}e^{-at}\sin(\omega t - \emptyset), \emptyset = tan^{-1}\dfrac{\omega}{-a}$	$\dfrac{1}{s[(s+a)^2 + \omega^2]}$
$1 - \dfrac{1}{\sqrt{1-\zeta^2}}e^{-\zeta w_n t}\sin(\omega_n\sqrt{1-\zeta^2}\,t + \phi), \quad \emptyset = cos^{-1}\zeta, \quad \zeta < 1$	$\dfrac{\omega_n^2}{s(s^2 + 2\zeta\omega_n s + \omega_n^2)}$
$\dfrac{\alpha}{a^2 + \omega^2} + \dfrac{1}{\omega}\left[\dfrac{(\alpha - a)^2 + \omega^2}{a^2 + \omega^2}\right]^{1/2}e^{-at}\sin(\omega t - \emptyset), \quad \emptyset = tan^{-1}\dfrac{\omega}{\alpha - a} - tan^{-1}\dfrac{\omega}{-a}$	$\dfrac{(s+\alpha)}{s[(s+a)^2 + \omega^2]}$

1.5. Dynamic Equations

We will take a small detour here to see the practical utility of mathematics we have studied and also slightly delve into the realm of writing "dynamic equations". This, we will do with the help of a practical example. Quite unconventionally, we will start with a problem statement.

A Practical Problem

A domestic, storage type water heater rated as 3 KW, 230 volts, 50 HZ, has a capacity of 3 litres and is seen to be working in steady- state with following observations:

The mass-flow rate of water = 0.03 kg/sec

The inlet temperature of water = 27°C

The outlet temperature of water = 48°C

The ambient temperature = 27°C

If there is a step change in the mass-flow rate from 0.03 kg/s to 0.05 kg/s, what will be steady state temperature of the water leaving the water heater and how much time will it take to attain a steady-state?

Note that there are two aspects in the problem: (i) steady-state detail, and (ii) Response speed. These two, in some sense, are related to the complimentary function and particular integral that we studied in the context of Linear Differential Equations.

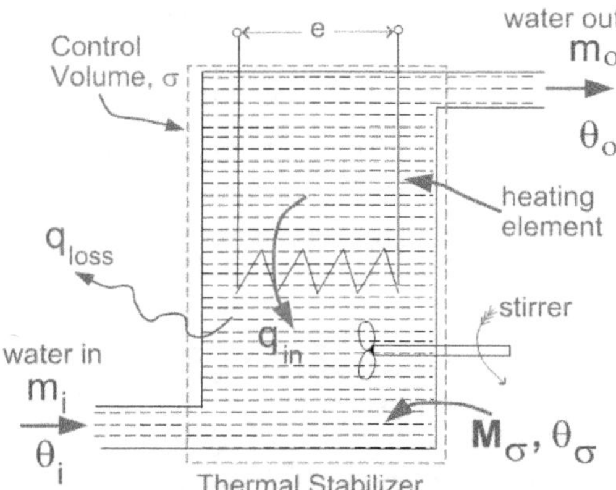

In problems where there are multiple conditions, we assume that one quantity is changing and others remain constant. For example, in the given problems, we assume that, when the inlet flow rate changed, the voltage, temperature of inlet water, and ambient temperature remain unchanged.

We present the mathematical modelling first and then show the "<u>Mathematicians approach</u>" and "<u>Control Engineer's Approach</u>". To begin the modelling of the system, let us consider the <u>schematic diagram</u> of the water heater.

Let us use the following terminology.

m_i, m_o = mass inflow and mass outflow rate (kg/sec)

σ = subscript used to indicate control volume

M_σ = Mass in control volume (kg)

θ_i, θ_o = inlet and outlet water temperatures (°C)

θ_σ = Temperature of water in control volume (°C)

q_{in} = Heat input (J/sec) or rating of heater

A = Exposed surface area of control volume (m^2)

h = convection coefficient $\left(\dfrac{J}{m^2 - °C - sec}\right)$

θ_a = ambient temperature (°C)

q_{loss} = heat loss $\left(\dfrac{J}{sec}\right)$

c_p = specific heat of water $\left(\dfrac{J}{Kg - °C}\right) = 4187 \dfrac{J}{kg - °C}$

Though these appear to be a dozen quantities, really we are dealing with five quantities: <u>mass, area, temperature, energy</u> and how some of these are changing with <u>time</u>. Note the units of these quantities.

If we carefully look at the phenomena, there are FIVE "processes" going in any time interval.

- Water is bringing in energy into control volume (Enthalpy in)
- Heating element is pumping heat in (q_{in})
- Temperature of water in control volume increases
- Water is taking out energy from control volume (enthalpy out)
- Energy is leaking from control volume (q_{loss})

Thus, in any time interval Δt seconds,

$$\overbrace{\left\{\begin{matrix}\text{Enthalpy}\\\text{in}\end{matrix}\right\} + \left\{\begin{matrix}\text{heat}\\\text{in}\end{matrix}\right\}}^{\text{cause}} = \overbrace{\left\{\begin{matrix}\text{Enthalpy}\\\text{out}\end{matrix}\right\} + \left\{\begin{matrix}\text{Change}\\\text{in interval energy}\end{matrix}\right\} + \left\{\begin{matrix}\text{heat}\\\text{loss}\end{matrix}\right\}}^{\text{effect}}$$

$$\Rightarrow (m_i c_p \theta_i)(\Delta t) + q_{in}(\Delta t) = (m_0 c_p \theta_0)(\Delta t) + M_\sigma c_p (\Delta \theta_\sigma) + q_{loss}(\Delta t)$$

$$\Rightarrow (m_i c_p \theta_i) + q_{in} = m_0 c_p \theta_0 + q_{loss} + (\Delta t) + M_\sigma c_p \frac{d\theta_\sigma}{dt} \qquad [as\ \Delta t \to 0]$$

Assumption 1:- The tank is always full. So, $m_i = m_0 = m$ (say)

$$\therefore\ mc_p \theta_i + q_{in} = mc_p \theta_0 + q_{loss} + M_\sigma c_p \frac{d\theta_\sigma}{dt}$$

Assumption 2:- Heat loss is predominantly convective: $q_{loss} = hA(\theta_\sigma - \theta_a)$

$$\therefore\ mc_p \theta_i + q_{in} = mc_p \theta_0 + hA(\theta_\sigma - \theta_a) + M_\sigma c_p \frac{d\theta_\sigma}{dt}$$

Assumption 3:- Water in the tank is well mixed. $\theta_\sigma = \theta_0$

$$\therefore\ mc_p \theta_i + q_{in} = mc_p \theta_0 + hA(\theta_0 - \theta_a) + M_\sigma c_p \frac{d\theta_0}{dt} \qquad (1)$$

This equation is the governing equation of the heater. We note here that when the heater is in <u>steady-state,</u> no quantity varies thus $\frac{d\theta_0}{dt} = 0$. Then

$$mc_p \theta_i + q_{in} = mcp_\theta + hA(\theta_0 - \theta_a) \qquad (2)$$

This equation is called <u>Steady Flow Energy Equation</u> (SFEE).

We use equation (1) and (2) to solve the problem.

Mathematician's Solution

The observations given for the <u>steady state</u> are:

m	= 0.03kg/sec	and q_{in}=3000 J/sec
θ_i	= 27°C	
θ_0	= 48°C	
θ_a	= 27°C	

Therefore using <u>SFEE,</u> we have

$$mc_p \theta_i + q_{in} = mc_p \theta_0 + hA(\theta_0 - \theta_a)$$

$$\therefore\ hA = \frac{mc_p(\theta_i - \theta_0)}{(\theta_0 - \theta_a)} = \frac{\{0.03 * 4187(27 - 48)\} + 3000}{(48 - 27)}$$

$$or\ hA = 17.25 \frac{J}{sec - °C}$$

Notice that the Steady State Values are Specified to Enable us to Compute the Value of hA.

To describe the "trajectory" along which the temperature changes with time when there is a disturbance, we use the Equation (1):

$$mc_p\theta_i + q_{in} = mc_p\theta_0 + hA(\theta_\sigma - \theta_a) + Mc_p\frac{d\theta_0}{dt} \qquad (1)$$

We will start measuring the time from the instant there is change: that is from the instant inlet water flow-rate changes. We say that $t=0$ when change occurred. When $t = 0, \theta_0 = 48°C$. When t= 0⁻ (just before the change occurred), m= 0.03 kg/sec. For all $t \geq 0$, the mass flow-rate m=0.05 kg/sec. Other than this change, all the other quantities remain at their values. The condition: "$t = 0, \theta_0 = 48°C$" is initial condition. From this initial condition, θ_0 varies, satisfying Equation (1). Our task is to find an expression for $\theta_0(t)$ as a function of time. For this, we substitute m=0.05kg/sec,

$$c_p = 4187\frac{J}{kg-0_C},\ \theta_i = 27°C,\ q_{in} = 3000\frac{J}{sec},\ hA = 17.25\frac{J}{sec-°C},\ \theta_a = 27°C, M_\sigma = 3kg$$

Into (1) and solve it with initial condition:$t = 0,\ \theta_0 = 48°C$.

Thus the equation is written as:

$$(0.05 * 4187 * 27) + (3000) = (0.05 * 4187)\theta_0 + 17.25(\theta_0 - 27) + 3 * 4187\frac{d\theta_0}{dt}$$

i.e. $5652.45 + 3000 = 209.35\theta_0 + 17.25\theta_0 - 465.75 + 12561\frac{d\theta_0}{dt}$

i.e. $12561\frac{d\theta_0}{dt} + 226.6\theta_0 = 9118.2$

To solve this equation, take Laplace transform on both sides with the notation that $£[\theta_0(t)] = \Psi(s)$ to get:

$$12561£\left[\frac{d\theta_0}{dt}\right] + £[226.6\theta_0] = £[9118.2]$$

$$i.e.,\ 12561[s\Psi(s) - 48] + 226.6\Psi(s) = \frac{9118.2}{s}$$

$$\left[\text{Recall that } £\left\{\frac{d\theta_0}{dt}\right\} = s£\{\theta_0\} - \theta_0|_{t=0} \quad and \quad £\{1\} = \frac{1}{s}\right]$$

$$\therefore (12561s + 226.6)\Psi(s) = \frac{9118.2}{s} + 602928 = \frac{9118.2 + 602928s}{s}$$

$$\Rightarrow \Psi(s) = \frac{602928s + 9118.2}{s[12561s + 226.6]} = \frac{A}{s} + \frac{B}{12561s + 226.6}$$

$$i.e.,\quad A(12561s + 226.6) + Bs = 602928s + 9118.2$$

$$Put\ s = 0: A(226.6) = 9118.2 \Rightarrow A = \frac{9118.2}{226.2} = 40.24°C$$

$$Put\ s = -\frac{226.6}{12561}: B\left(-\frac{226.6}{12561}\right) = 602928\left(-\frac{226.6}{12561}\right) + 9118.2 = -1758.6$$

$$\therefore B = 97483.5°C$$

$$\therefore \Psi(s) = \frac{40.24}{s} + \frac{97483.5}{12561s + 226.6} = \underbrace{\frac{4024}{s}}_{\text{Steady state part}} + \underbrace{\frac{7.76}{s + 0.018}}_{\text{Transient part}}$$

Now take inverse transform to get

$$\theta_0(t) = 40.24 + 7.76e^{-0.018t}$$

As a verification, check what is θ_0 at $t = 0$

For steady state, $t \to \infty$ $\therefore \theta_0 = 40.24°C$

We specify <u>2% settling time</u>, t_s, as time required to attain steady state.

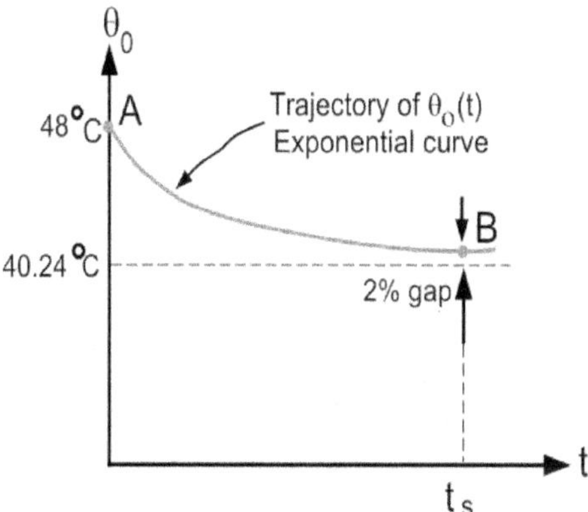

At point B: $\theta_0(t) = 40.24 + 2\%$ of 40.24

or, $40.24 + 7.76e^{-0.018t_s} = 41.0448 \Rightarrow t_s \approx 126\ sec$

$\therefore$ The new steady state temperature $\theta_0 = 40.24°C$ is attained 126 sec after the change occurred.

Control Engineer's Solution

The observations given for steady state are:

m	= 0.03kg/sec	and q_{in}=3000 J/sec
θ_i	= 27°C	
θ_0	= 48°C	
θ_a	= 27°C	

Therefore, using SFEE:

$$mc_p\theta_i + q_{in} = mcp_\theta + hA(\theta_0 - \theta_a)$$

$$\therefore hA = \frac{mc_p(\theta_i - \theta_0)}{(\theta_0 - \theta_a)} = 17.25 \frac{J}{sec - °C}$$

This part is same as the mathematician approach.

At the new <u>steady state,</u> $m = 0.05 \frac{kg}{sec}$, $\theta_i = 27°C$, $\theta_a = 27°C, q = 3000\frac{J}{sec}$.

$$(0.05 * 4187 * 27) + 3000 = (0.05 * 4187)\theta_0 + 17.25(\theta_0 - 27)$$

$$\Rightarrow \theta_0 = \frac{9118.2}{226.6} = 40.24°C \quad \text{[Fortunately same as we got earlier.]}$$

To find the time taken to attain steady state, start with equation (1)

$$mc_p\theta_i + q_{in} = mc_p\theta_0 + hA(\theta_\sigma - \theta_a) + mc_p\frac{d\theta_0}{dt}$$

Apply difference operator on either side and note that $\Delta(xy) = (\Delta x)y + x(\Delta y)$ to get:

$$\Delta(m)c_p\theta_i + mc_p(\Delta\,\theta_i) + \Delta\,q_{in} = (\Delta m)c_p\theta_0 + mc_p(\Delta\theta_0) + hA(s\theta_0 - \Delta\,\theta_a + M_\sigma c_p\frac{d(\Delta\theta_0)}{dt}$$

Note that θ_i did Not change $\therefore \Delta\theta_i = 0$; q_{in} did Not change $\therefore \Delta q_{in} = 0$ ambient temperature did Not change $\therefore \Delta\theta_a = 0$

$$\Rightarrow (\Delta m)c_p\theta_i = (\Delta m)c_p\theta_0 + [(mc_p + hA) + m_\sigma c_p D]\theta_0 \text{ where } D \triangleq \frac{d}{dt}$$

$$\Rightarrow \frac{\Delta\theta_0}{\Delta m} = \frac{c_p(\theta_i - \theta_0)}{(mc_p + hA) + M_\sigma c_p D} = \frac{c_p(\theta_i - \theta_0)/mc_p + hA}{1 + \frac{M_\sigma c_p}{mc_p + hA}D} = \frac{c_1}{1 + \tau D}$$

where $c_1 = \frac{c_p(\theta_i - \theta_0)}{(mc_p + hA)}$ and $\tau = \frac{M_\sigma c_p}{(mc_p + hA)} =$ time constant.

$$\tau = \frac{3 * 4187}{0.03 * 4187 + 17.25} \cong 88\ sec$$

It is known that 2% settling time for a first order system is $\approx 4\tau = 352\ sec.$

Which solution did you like? Why?

The two approaches gave same new steady-state temperature but different 2% settling times. Which one should we accept as correct? Obviously mathematicians approach is accurate but in a bit more complicated situation, mathematicians approach may pose lot more complexity. Also, control engineers approach overestimated the settling time. So, a system designed as per control engineers' approach in this case will prove to be faster than the settling time for which it was designed.

1. To conclude it all, this lecture attempts to bring forth a few more important concepts into fore front:
2. <u>Modelling a system,</u> i.e., writing dynamic equations of the system being studied, is a important step in analyzing and designing control systems.
3. In the context of Linear systems, <u>good understanding of Laplace transform is essential</u>.
4. Very many times, instead of dealing with absolute quantities, it may be easier to deal with changes using Δ operator. This process in a very general sense may be called <u>linearizing</u>.
5. The concept of transfer function becomes very handy since all known results about known transfer function can be readily applied.
6. Sometimes, an approximation gives reasonably "good" solutions but sometimes NOT.
7. One need NOT be biased either towards mathematicians approach or should NOT oppose either one. Both have their own appeal and use. Hence we need to learn both!

Home Work 06

A domestic, storage type water heater rated as 3 KW, 230 volts, 50 HZ, has a capacity of 3 litres and is seen to be working in steady- state with following observations:

> The mass-flow rate of water = 0.03 kg/sec
>
> The inlet temperature of water = 27°C
>
> The outlet temperature of water = 48°C
>
> The ambient temperature = 27°C

While it is running at steady-state, a voltage surge occurred and the voltage raised to 260V. This surged voltage remained at 260 V for sufficiently long time duration. What will be new steady state temperature of the water leaving the water heater and how much time will it take to attain a steady-state?

1.6. Transfer Function

We have seen one example problem to show the analysis procedure and as an alternative method, it is shown that an approximate and simplified, general method is possible using ratio of output (or effect) to input, termed "Transfer function (TF)". In fact the concept of transfer function is central to finite dimensional, linear, time, invariant (LTI), continuous (or discrete time) systems and most of our course revolves around it.

In subsequent discussion: (i) we define TF, (ii) Show how to obtain TF of mechanical spring, mass, damper (SMD) systems, (iii) RLC circuits, (iv) Analogous relation between SMD and RLC, (v) TF of feedback systems, (vi) Block diagram, signal flow graph, (vii) Armature controlled DC servomotor, field controlled DC servomotor, (viii) AC motors.

The list is kind of long, but we begin by taking one step at a time, as it should be done. The definition of TF is given below:

Definition:- Transfer Function is the ratio of Laplace transform of output to the Laplace Transform of input of LTI Single-Input-Single-Output (SISO) system with the assumption that all initial conditions are zeros[2].

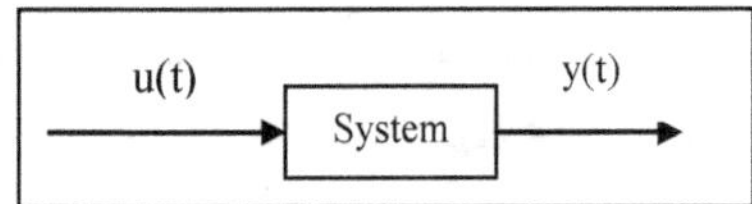

Black Box Representation of System

Consider a Single-Input-Single-Output (SISO), Linear time-invariant (LTI) continuous time system with the u(t) as input and y(t) as output. Since the system is LTI system, it is possible to write <u>dynamic</u> relation between the input and output as[3]:

$$a_n \frac{d^n y}{dt^n} + a_{n-1} \frac{d^{n-1} y}{dt^{n-1}} + \cdots + a_1 \frac{dy}{dt} + a_0 y = b_m \frac{d^m u}{dt^n} + b_{n-1} \frac{d^{m-1} u}{dt^{m-1}} + \cdots + b_1 \frac{du}{dt} + b_0 u \quad ----(A)$$

with $a_n \neq 0, b_n \neq 0$. Let $Y(s) = £\{y(t)\}$ and $U(s) = £\{u(t)\}$ and take Laplace Transforms on both sides of the dynamic relation to get:

$$(a_n s^n + a_{n-1} s^{n-1} + \cdots + a_1 s + a_0) Y(s) = (b_m s^m + b_{n-1} s^{m-1} + \cdots + b_1 s + b_0) U(s)$$

Carefully apply theorem on Laplace Transform of derivative of a variable and simplify to convince yourself that the last equation above is correct when all initial conditions are zeros[4].

$$\therefore \frac{Y(s)}{U(s)} = \frac{b_m s^m + b_{m-1} s^{m-1} + \cdots + b_1 s + b_0}{a_n s^n + a_{n-1} s^{n-1} + \cdots + a_1 s + a_0} = G(s) = \frac{N(s)}{D(s)} \quad ------------- (B)$$

Note

1. The order of TF is the order of denominator polynomial.
2. Roots of denominator of TF are called <u>Poles</u> and the roots of the numerator are called <u>Zeros</u>.

---

3. TF has units since input and output of a physical system have units E.g.: if u(t)is in Newtons and y(t) is in meters, G(s) has units of meters/Newton.

4. Let u (t) = unit impulse input. Then U(s) =1.

   ∴ From Equation (B), we can obtain Laplace transform of output as:

$$Y(s) = \frac{b_m s^m + b_{m-1} s^{m-1} + \cdots + b_1 s + b_0}{a_n s^m + a_{n-1} s^{n-1} + \cdots + a_1 s + a_0} = G(s)$$

$$\therefore y(t) = £^{-1}[Y(s)] = g(t) = £^{-1}[G(s)] \text{-----------(C)}$$

g (t) in equation (C) is called <u>impulse Response</u>. **Thus, TF may also be defined as the Laplace transform of impulse Response of the system with zero initial conditions.**

1. D(s) =0 is termed the <u>Characteristic equation</u> of the system. Roots of D(s) = 0 indicate the <u>Transient response</u> and the <u>stability</u> characteristics of the system.

2. For <u>Casual systems</u>, n≥ m, why?

   The transfer function G(s) is Strictly proper if n>m.

   G(s) is called <u>Proper if n=m</u>. G(s) is termed <u>improper if n<m</u>. n is <u>order</u> of TF.

[5]We now try to write <u>Dynamic equations</u> of a few simple SMD and RLC systems and derive TF of these.

Example 1

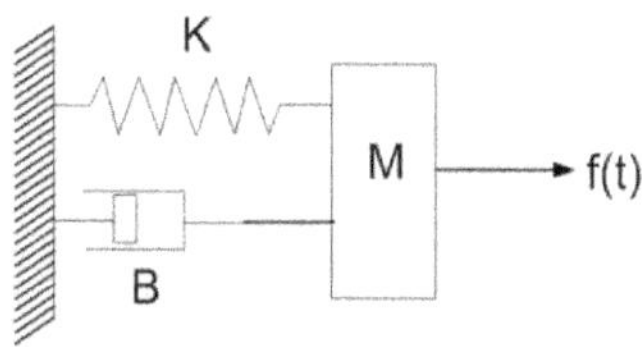

For the spring, mass, damper system shown in figure, derive the transfer function relating x to f where f(t) is the exciting force in Newtons, x(t) is the displacement in meters, k is the spring stiffness in N/m, and b is the damping coefficient in N-s/m. The position of mass M (in kilograms) is such that at t=0, it is at rest in response position, x=0

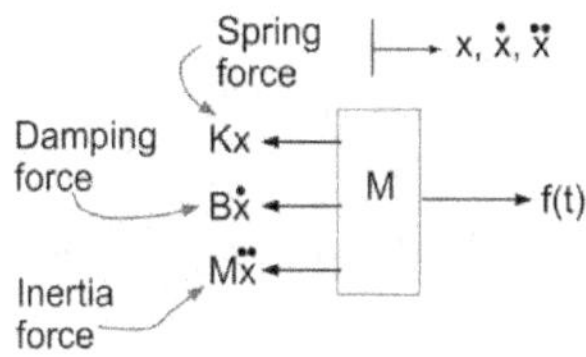

Solution

We need to write dynamic equations from <u>first principles</u>. In this case, we invoke <u>Newton's Second law</u> for the mass M. To that end we draw the Free body Diagram (FBD) of M.

A force balance gives dynamic equation as:

$$f(t) = m\ddot{x} + B\dot{x} + kx$$

Using the notation $x(s) \triangleq £\{x(t)\}$ and $F(s) \triangleq £\{f(t)\}$ and taking Laplace transforms on both sides, we get,

$$F(s) = (Ms^2 + Bs + K)X(s)$$

$$\therefore \frac{x(s)}{F(s)} = \frac{1}{Ms^2 + Bs + k}$$

> Don't go beyond this point if you don't understand this.

This is a second order <u>proper transfer</u> function. We study about the response of standard second order system at a later part of the course.

What are the units of the transfer function derived?

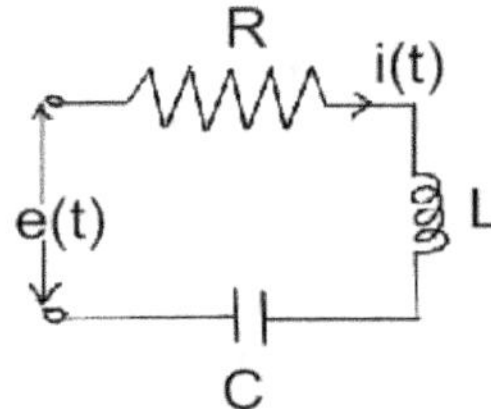

Example 2

For the RLC circuit shown in the figure, derive the TF relating i(t) to e(t)

Solution

We start with Kirchhoff's voltage law to write:

$$e(t) = Ri(t) + L\frac{di(t)}{dt} + \frac{1}{C}\int_0^t i(\tau)d\tau$$

Taking Laplace transforms on both sides under zero initial conditions, we get[6],

---

[6] Is this a proper transfer function? Does it have a zero? What are the units of this TF? See (ii) on Page 2.

$$E(s) = \left[R + LS + \frac{1}{Cs}\right]I(s) \Rightarrow \boxed{\frac{I(s)}{E(s)} = \frac{Cs}{Lcs^2 + RCs + 1}}$$

Do Not go beyond this point until you understand everything very clearly and you can reproduce effortlessly the essence of what you learnt so far.

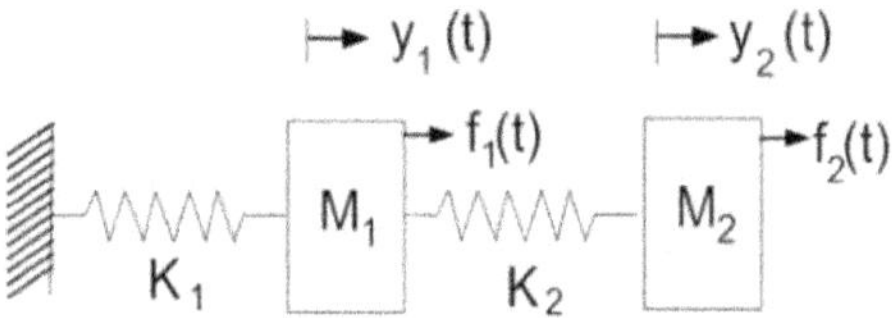

Example 3

Find the Transfer function $\frac{y_2s}{F_1s}$ for the spring, Mass system.

Solution

There are two input forces: $f_1(t)$ and $f_2(t)$. When we want to find $\frac{Y_2(s)}{F_1(s)}$, we set $f_2(2)= 0$. Setting $f2(t)= 0$, we draw FBD of M_1 and M_2 as shown below:

∴ The dynamic equations may be written as:

$$f_1(t) = M_1\ddot{y}_1 + K_2(y_1 - y_2) + K_1 y_1$$
$$0 = M_2\ddot{y}_2 + K_2(y_2 - y_1)$$

Taking Laplace transforms on either side of each equation with zero initial conditions and simplifying, we get:

$$\frac{Y_2(s)}{F_1(s)} = \frac{K_2}{M_1 M_2 s^4 + [K_2 M_1 + \{K_1 + K_2\}M_2]s^2 + K_1 K_2}$$

Derive this and convince yourself that this is the true answer.

Home Work 07

(Q1).   Find the transfer function $\frac{x(s)}{F(s)}$ for the system shown in the figure. M is the mass of block in kilograms k is the stiffness on spring in N/m, x(t) is vertical displacement of mass in meters, and f(t) is the force acting on mass M. Assume zero initial conditions.

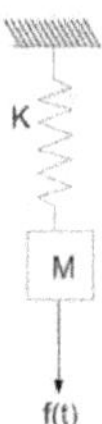

(Q2).   For the RLC circuit shown in the figure below, find the TF $\frac{E_o(s)}{E_{in}(s)}$

What are the poles, zeros of this transfer function? Is it strictly proper, or proper, or improper transfer function? What is the time constant of the system?

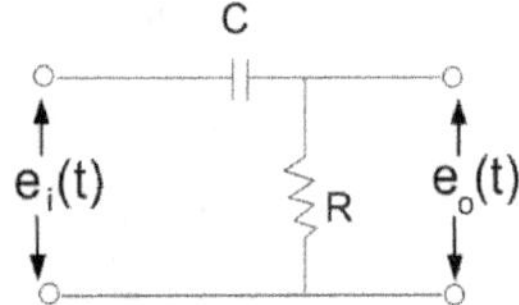

(Q3).   For the mechanical system shown, find the Transfer function $\frac{Y_1(s)}{F(s)}$. f(t) is the input force and y_1, y_2 are output displacements.

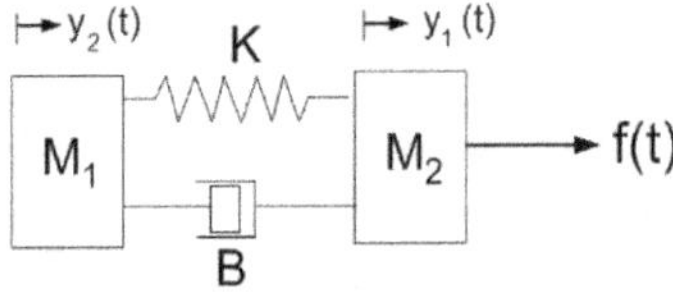

(Q4).   Find the TF for the following circuit

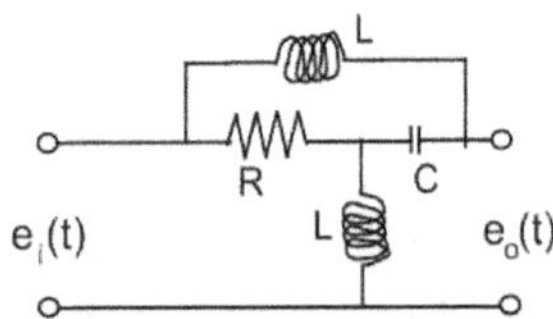

1.7.   Mechanical Torsional Systems

In many problems dealing with motors, engines, rotary compressors and fans, we come across rotating elements with inertia or torsional stiffness or torsional viscous damping. Though the analysis is much similar to the linear, Translatory Mechanical Systems, torsional systems are briefly discussed here for continuity.

In studying Mechanical Torsional Systems, the parameters we study are inertia, torsional stiffness, and torsional viscous damper. The physical principle is Newton's Second Law here too.

Table below shows Torsional system elements and corresponding Translatory system elements.

	Torque T (N-m) Vector	Polar Moment of inertia J (kg-m²)	Angular Displacement θ (radians) Vector	Angular Velocity $\dot{\theta}$ (radians/s) Vector	Angular Acceleration $\ddot{\theta}$ (rad/s²) Vector	Torsional spring K_T $\left(\dfrac{N-m}{rad}\right)$	Torsional damper b_t (n-m-s/rad)
Mechanical Torsional System		Scalar $T = J\dfrac{d^2\theta}{dt^2}$				Scalar $T = K_L\theta$	Scalar $T = b_t\dot{\theta}$
Mechanical Translator system	Force F (N) Vector	Mass M (Kg) Scalar $F = m\dfrac{d^2x}{dt^2}$	Linear displacement X (m) vector	Linear velocity $\dot{x}$ (m/s) vector	Linear acceleration $\ddot{x}$ (m/s²) Vector	Linear spring K (n/m) Scalar $F = Kx$	Linear damper b (n-m-s/m) Scalar $F = b\dfrac{dx}{dt}$

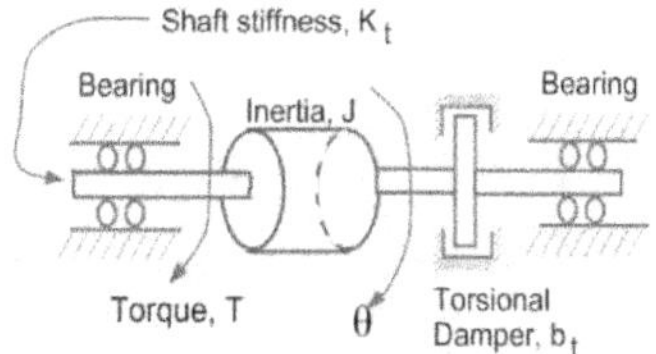

We present an example to illustrate the information present in the table above.

Example 1

Find the transfer function relating θ to T

Solution

The torque moves the inertia, twists the shaft, and turns the damper. Therefore, we write

$$T = J\frac{d^2\theta}{dt^2} + b_t\frac{d\theta}{dt} + K_t\theta$$

Notice that T(t), $\theta(t)$ are functions of time. Taking Laplace Transforms on either side with zero initial conditions we get

$$T(s) = Js^2\theta(s) + b_t s\theta(s) + K_t\theta$$
$$T(s) = (Js^2 + b_t s + K_t)\theta(s)$$
$$\therefore \frac{\theta(s)}{T(s)} = \frac{1}{J s^2 + b_t s + k_t}$$

Example 2

A motor, with total inertia of J_m, moves a pump with total inertia of J_L. The total viscous damping on the <u>motor side</u> is b_m and that on the pump side (or <u>load–side</u> or is b_2. Write the dynamic equations describing the system.

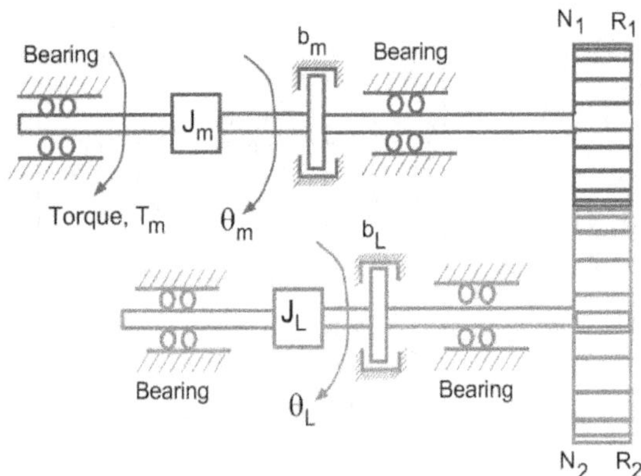

Solution

Let F_t be the tangential force at the pitch circle. Then, the equations may be written as:

$$T_m = J_m\ddot{\theta}_m + b_m\dot{\theta}_m + F_t R_1$$
$$F_t * R_2 = J_L\ddot{\theta}_L + b_L\dot{\theta}_L$$

Home Work 7a

(Q1). Find the transfer function $\frac{\theta_m(s)}{T_m(s)}$ for the system shown in Example 2.

1.8.   Electro Mechanical Analogy

This little module is introduced here to give a feel of the "links" that exist in the nature of dynamics of "electrical circuits" and "mechanical circuits."

Mechanical systems wherein movement/deflection/force and its implications are of importance are often modeled using elements that exhibit stiffness, elements that exhibit inertia and elements that exhibit energy dissipation. Though in general such elements do not appear explicitly as a spring, a damper, and a lumped mass, to facilitate visual inspection and understanding of the interconnections, we use spring(⌇⌇⌇) to represent stiffness, mass($\boxed{m}$) to represent inertia and viscous damper(⌑)to represent energy dissipation. Such representation allows us to arrive at lumped parameter models which are amenable to analysis using ordinary differential equations.

Experimenting with lumped parameter mechanical models by actually building systems using spring, mass, damper (SMD), takes more effort than building electrical circuits using resistance, inductance, and capacitance. This fact prompts us to study analogous systems of SMD and RLC. We say that two systems are analogous if their transfer functions appear in similar form mathematically, or their transfer functions are identical.

Consider the SMD circuit and the RLC circuit shown below:

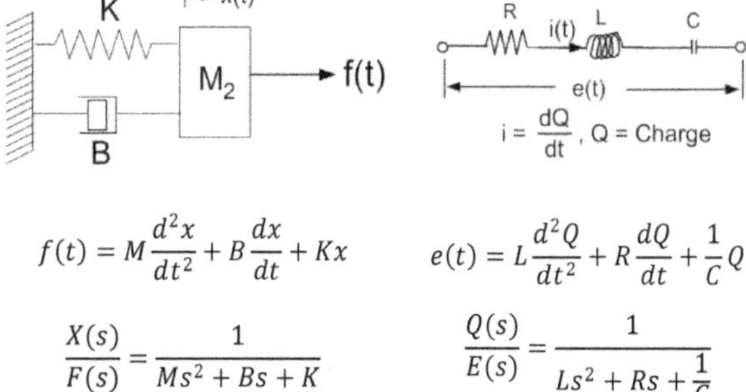

$$f(t) = M\frac{d^2x}{dt^2} + B\frac{dx}{dt} + Kx \qquad e(t) = L\frac{d^2Q}{dt^2} + R\frac{dQ}{dt} + \frac{1}{C}Q$$

$$\frac{X(s)}{F(s)} = \frac{1}{Ms^2 + Bs + K} \qquad \frac{Q(s)}{E(s)} = \frac{1}{Ls^2 + Rs + \frac{1}{C}}$$

Note that, if we choose the numerical values of parameters of electrical system as: L = M, R = B, and $\frac{1}{C} = K$, the transfer functions of both the systems shown are identical. thus, the SMD-circuit and RLC-circuit are analogous.

We identify the force-voltage analogy using the analogous quantities given in table below. This analogy is often termed "Direct Analogy."

SMD-Circuit	Force, f	Mass, M	Damper, B	Spring, K	Displacement, x	Velocity, $\frac{dx}{dt}$
RLC-Circuit	Voltage, e	Inductance, L	Resistance, R	Capacitance, $\frac{1}{C}$	Charge, Q	Current, i

Notice that, in the mechanical system, M, K, and B are in parallel whereas R,L, and C in electrical system are in series. Thus, the series elements in SMD-circuit will be in parallel in analogous RLC-circuit and vice versa.

1. Mass, spring, and damper undergo same deflection in SMD-circuit. Inductance, Capacitance, and Resistance carry same current.

2. Sum of inertia force, damping force, and spring force is equal to the external force in SMD-circuit. Sum of voltage drops across inductance, resistance, and capacitance is equal to external voltage, e, in the analogous RLC-Circuit.

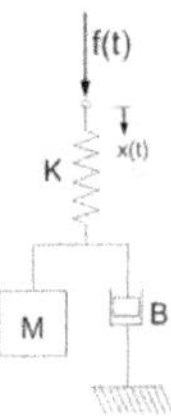

Example: Electrical Analog

The mechanical system shown in figure beside needs to be studied using an electrical analog. Specify the layout and values of R, L, and C.

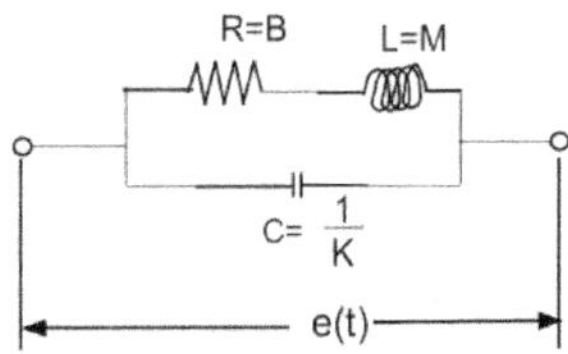

Solution: Notice that M and B are parallel in the SMD system. Hence, L and R need to be in series in electrical analog. Similarly M-B combination is in series with K in SMD system. Hence R-L combination needs to be in parallel with the capacitance. Noting this, the electrical analog may be drawn as shown in Figure beside. The numerical values of the resistance, inductance, and capacitance in the electrical analog may be taken as :R = B, L = M, and $C = \frac{1}{K}$. The student is advised to derive the transfer function and verify that the two systems are indeed analogous.

Scale Factors

Suppose that in the example considered just now, the force is of the order of 1000 N and the velocity of mass M is of the order of 10 m/s. Then, the voltage in electrical analog needs to be of the order of 1000 V and the current needs to be of the order of 10 A. Such high voltages and currents are very inconvenient to handled and not safe too. To obviate such inconvenience, we can use "scale factors" much the same way we do in the case of drawing a map or drawing a graph.

Let K_e be the voltage scale factor and K_i be the current scale factor so that

$$f = K_e e \text{ and } \frac{dx}{dt} = K_l I.$$

With the choice of the scale factors, we need to adjust the values of R, L, and C to meet this specification.

Equation for SMD-Circuit	Equation for RLC-Circuit	
$f = M\dfrac{d}{dt}\left(\dfrac{dx}{dt}\right)$	$E = L\dfrac{d}{dt}I$	Row 1: Mass-inductance
$f = B\dfrac{dx}{dt}$	$E = RI$	Row 2: Damper-Resistance
$f = Kx \Rightarrow \dfrac{df}{dt} = K\dfrac{dx}{dt}$	$\dfrac{dE}{dt} = \dfrac{1}{C}I$	Row 3: Spring-Capacitance

Consider row 1 and write $\dfrac{f}{E} = \dfrac{M\frac{d}{dt}\left(\frac{dx}{dt}\right)}{L\frac{d}{dt}I} = \dfrac{M\frac{d}{dt}(K_iI)}{L\frac{d}{dt}I} = \dfrac{M}{L}K_i$. Hence $L = M\dfrac{K_i}{K_e} = rM$ where $r = \dfrac{K_i}{K_e}$.

Simplifying from row 2, $\dfrac{f}{E} = \dfrac{B\frac{dx}{dt}}{RI} \Rightarrow R = rB.$

And for row 3, $\dfrac{\frac{df}{dt}}{\frac{dE}{dt}} = \dfrac{K\left(\frac{dx}{dt}\right)}{\frac{1}{C}I} \Rightarrow C = \dfrac{1}{rK}.$

Example: Electrical analog with scale factors

The parameters of a mechanical system are M = 10 kg and B = 500 N-s/m, and K = 2500 N/m. It is estimated that f_{max}=5000 N and $\dot{x}_{max} = 5\frac{m}{s}$. Determine the values of resistor, capapcitor, and inductor to use in the direct analog such that E_{max} = 100 V and I = 4 A.

After the analog has been built and used in experiments, in a particular situation, it is measured that E = 50 V and I = 4 A. What were the actual values of f and $\dot{x}$?

<u>Solution:</u> $f_{max} = K_e E_{max} \Rightarrow K_e = \dfrac{f_{max}}{E_{max}} = \dfrac{5000}{100} = 50\dfrac{N}{V}.$

Similarly, $\dot{x}_{max} = K_i I_{max} \Rightarrow K_i = \dfrac{\dot{x}_{max}}{I_{max}} = \dfrac{5}{10} = \dfrac{1}{2}\dfrac{m/s}{Amp}.$

Hence, $r = \dfrac{K_i}{K_e} = \dfrac{1}{100}$. Therefore,

$$L = rM = \frac{1}{100}\,10 = 0.1H$$

$$C = \frac{1}{rK} = \frac{1}{\frac{1}{100}\,2500} = 0.04\ F$$

$$R = rB = \frac{1}{100}\,5000 = 50\Omega$$

In the experiment, if E = 50 V, then, $f = K_e E = 50 \times 50 = 2500\ N$ and

$$\text{if I = 4 A, } \dot{x} = K_i I = \frac{1}{2} \times 4 = 2\frac{m}{s}.$$

Time Scale

For mechanical systems which act very fast, it may be desirable to have the analog go a bit slow so that we can watch the variables closely. Similarly, when mechanical systems are very slow, we may want to have the electrical analog go faster so that we don't have to wait long to observer phenomena. To achieve such change in the time scale, we use a "time scale."-

Let

t = time taken for a phenomenon to occur in the mechanical system, and

τ = time taken for corresponding phenomenon to occur in electrical analog.

If these times are related by $\tau = at$, we note that $a > 1$ implies that electrical analog is slower than mechanical system and $a < 1$ implies that the electrical analog is faster than the mechanical system. With $\tau = at$, we write, $\frac{dt}{d\tau} = \frac{1}{a}$. To achieve the speeding up or slowing down depending on the vale of a, we need to further adjust the values of R, L, and C.

To derive the value of L, for example, we write:

$$\left. \begin{array}{l} E = L\dfrac{dI}{d\tau} = L\dfrac{dI}{dt}\dfrac{dt}{d\tau} = L\dfrac{dI}{dt}\dfrac{1}{a} \\[2mm] f = M\dfrac{d}{dt}\left(\dfrac{dx}{dt}\right) = M\dfrac{d}{dt}(K_i I) = MK_i\dfrac{dI}{dt} \end{array} \right\} \Rightarrow \begin{array}{l} \dfrac{E}{f} = \dfrac{1}{K_e} = \dfrac{L\frac{1}{a}}{MK_i} \\[3mm] \Rightarrow L = a \times \left(\dfrac{K_i}{K_e}\right)M = arM. \end{array}$$

Similarly, writing equations for resistor we can determine the values of resistor and capacitor and summarize,

$$L = arM$$

$$R = rB$$

$$C = \frac{a}{rK}$$

The student is encouraged to notice carefully that the value of R is not affected by the time scale. Why is this so?

Home Work 8

Q1)   For the mechanical system shown in Figure, construct a direct analog. Given M= 0.5 kg, B = 10 N-s/m, K = 50 N/m, K_e = 2 N/V, and K_i = 200 m/(s-A), if the solution needs to be slowed down in the analog by a factor of 10, specify the values of electrical elements.

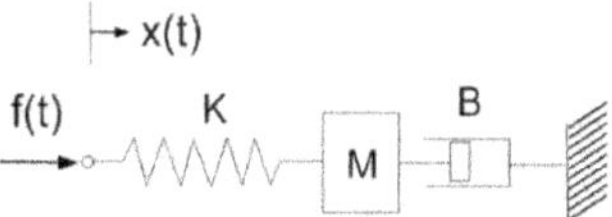

Q2)   Show that the two systems shown in Figure can be made analogous by properly choosing the transfer function and the values of parameters of the electrical circuit.

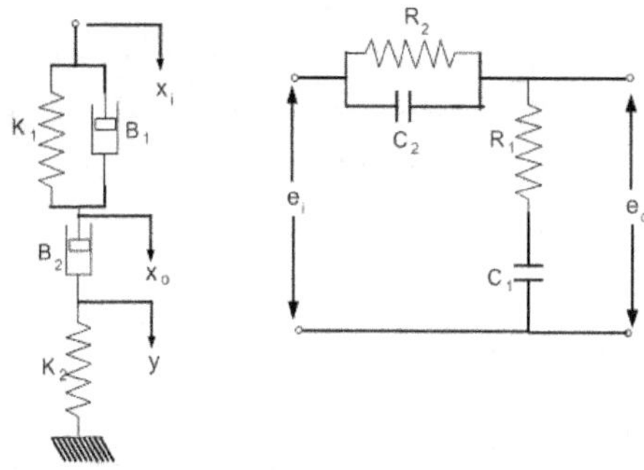

Q3)   Construct an electrical analog for the mechanical system shown beside.

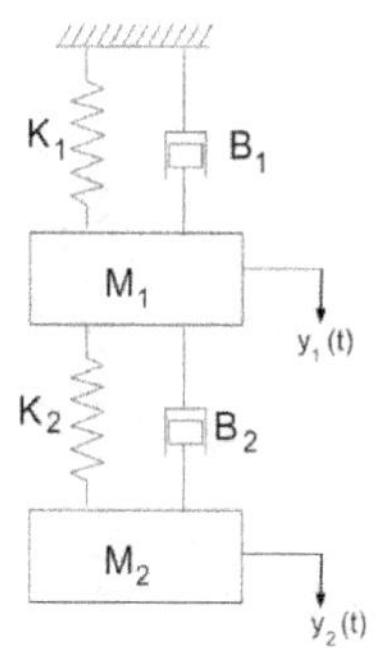

35

1.9.   DC – Servomotor

Broadly, a servomotor is a device/system used in controlling position or velocity or acceleration. Depending on the actuation, we may talk about hydraulic servomotor, pneumatic servomotor, or electric servomotor. In case of electric servomotors, we have AC servomotors or DC servomotors. The topic of this discussion, DC-servomotors will be dealt under two headings:

1. Armature controlled DC-servomotor wherein the output torque of the motor can be varied by varying the current flowing through Armature windings and felid current is held constant.

2. Field controlled DC-servomotor wherein output torque of the motor can be varied by varying the current flowing through the field and windings and armature current is held constant.

We study (i) & (ii) above as a part of this discussion both in open loop and closed loop configuration.

While we study these and try to develop these dynamics and extend the <u>Conceptual block diagram</u> we studied in NOTES – 01 into a <u>Mathematical block diagram</u> and obtain transfer function of the systems we study. Thus, this discussion serves us in revising TF – ideas and broach block diagrams Reduction ideas. First let us begin with (i) often funnily abbreviated as <u>AC DC servomotor.</u>

Armature Controlled DC– Servomotor

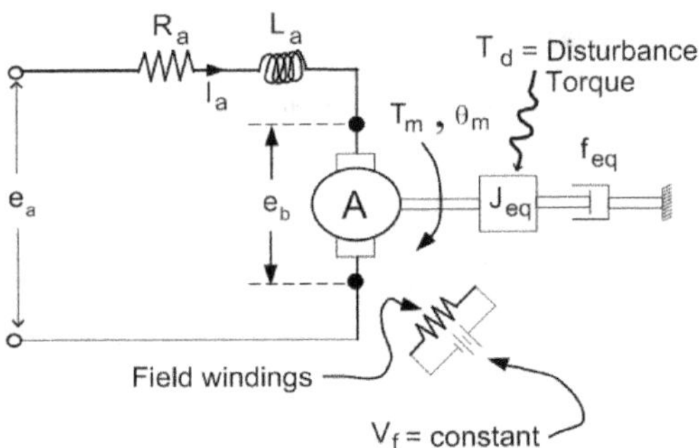

Figure beside shows schematic of the motor under consideration with separately excited felid windings under constant excitation voltage. The armature windings have a resistance of R_a and an inductance of L_a . When a <u>control voltage</u> of e_a is applied, the armature current of I_a

flows and, by DC–motor action, a torque of T_m is generated at motor shaft and the <u>equivalent inertia</u> J_{eq} and <u>equivalent damping</u>, f_{eq} get an the angular displacement of θ_m . The e_b shown in the schematic above is <u>back emf</u> which is directly proportional to speed of armature shaft and opposes control voltage[7].

We try to "break- up" the schematic into Conceptual block diagram, write mathematical equations for each block, and draw the overall block diagram to obtain transfer function. The complete block diagram is drawn as:

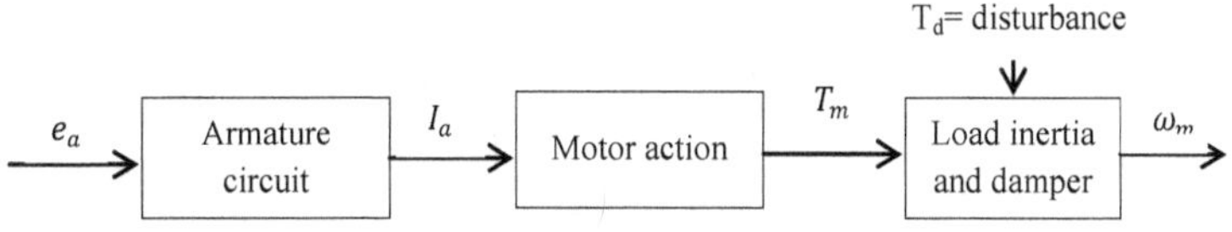

We now write equations for each block ion conceptual block diagram

Armature Circuit

$$\because e_a = e_b + (R_a + L_a D)I_a \quad ---(1)$$

$$D \triangleq \frac{d}{dt}$$

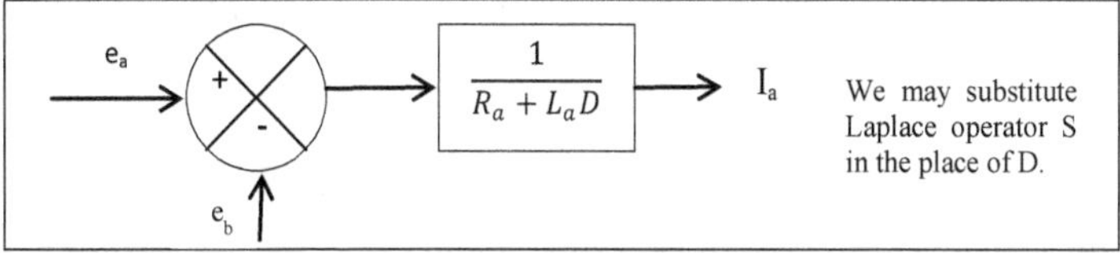

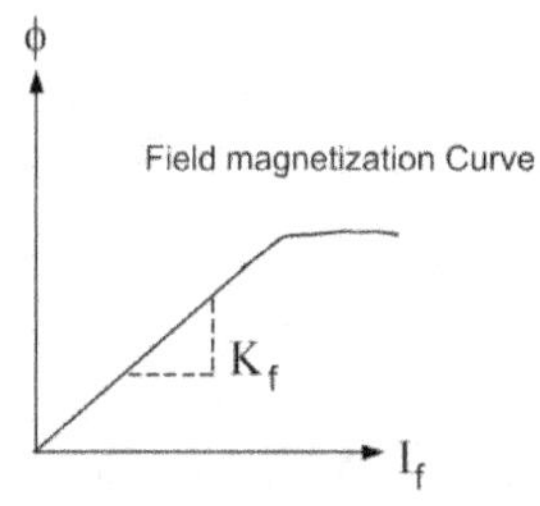

---

⁷ <u>Food for thought:</u>
- Why there is an inductor in "DC" servomotor?
- Is there a connection between e_b and 2nd Law of thermodynamics?
- Why do we call equivalent inertia and equivalent damping?

Motor Action

Torque produced is proportional to flux and is also proportional to armature current.

$$\therefore\ T_m \propto \emptyset\, I_a$$

The field magnetization curve shown saturates and becomes nonlinear beyond a range as shown in the Figure. If we consider linear range as our operating range,

$$\emptyset \propto I_f\ \therefore\ T_m \propto I_f I_a\,.$$

$$\text{Thus, } T_m = K\, I_f\, I_a = K_m\, I_a \qquad ---- (2)$$

$K_m = KI_f$ is called as motor torque constant.

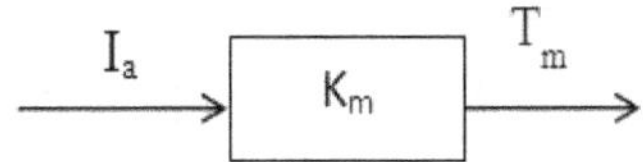

Load on Motor

The torque developed needs to oppose disturbance torque, move the inertia, and move the damper.

$$\therefore\ T_m = T_d + \left(J_{eq}\, D + f_{eq}\right)\omega_m \qquad -----(3)$$

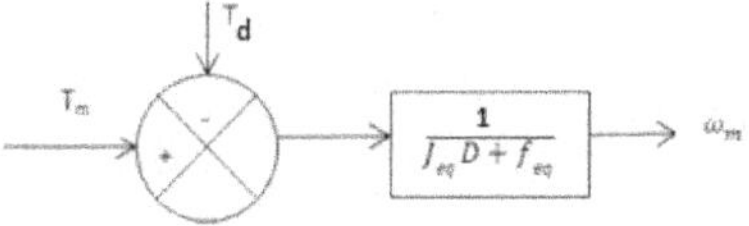

Back EMF Equation

$$e_b = K_b\,\omega_m \qquad -----(4)$$

Where K_b is back emf constant

From (1), (2), (3), (4), by simplification, we may obtain

$$\omega_m = \left[\frac{K_m}{(R_a + L_a D)(J_{eq} D + f_{eq}) + K_m K_b}\right] e_a - \left[\frac{(R_a + L_a D)}{(R_a + L_a D)(J_{eq} D + f_{eq}) + K_m K_b}\right] T_d \quad ----(5)$$

Two TFs are shown in above equation[8]: one acting on e_a and the other acting on T_d. The TF shown in the left-most box is TF from e_a to ω_m and the TF shown is TF from T_m to ω_m.

The overall mathematical block-Diagram may be drawn by joining individual block diagrams

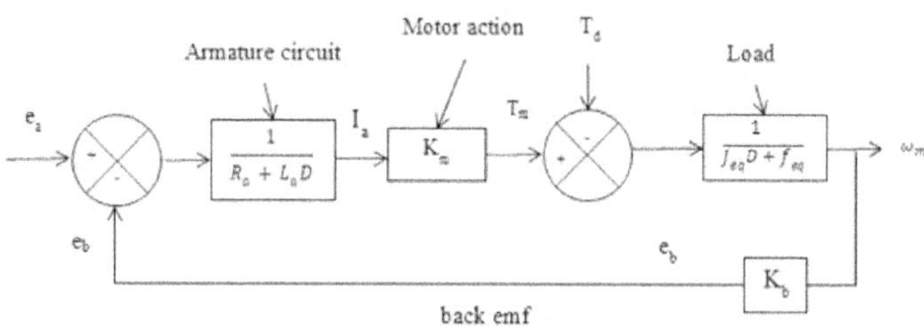

Usually L_a is very small $\Rightarrow L_a \approx 0$. So the equation (5) reduces to[9]

$$\omega_m = \frac{K_m}{R_a\left(J_{eq} D + f_{eq}\right) + K_{m^2}}\, e_a - \frac{\left(R_a + L_a \uparrow^0 D\right)}{R_a\left(J_{eq} D + f_{eq}\right) + K_{m^2}}\, T_d$$

Rearranging terms in above equation,

$$\omega_m = \frac{K_m \big/ \left(f_{eq} R_a + K_{m^2}\right)}{1 + \tau_0 D}\, e_a - \frac{R_a \big/ \left(f_{eq} R_a + K_{m^2}\right)}{1 + \tau_0 D}\, T_d ----(6)$$

Where

$$\tau_0 = \frac{J_{eq} R_a}{f_{eq} R_a + K_{m^2}} = \text{Time constant}$$

<u>Food for thought:</u> Systems with small time constant act faster. Think through and see under what conditions does τ_0 above give faster response?

Equation (6) gives transient behavior of the system whereas; rated capacities are usually specified for steady state conditions. To distinguish between transient and steady-state values, we use a subscript "ss" as in $\omega_{m,ss}$, $T_{m,ss}$

In Equation (6) if we substitute D= 0, we get steady equations. Why?

$$\therefore\ \omega_{m,ss} = \frac{K_m}{f_{eq} R_a + K_{m^2}}\, e_a - \frac{R_a}{f_{eq} R_a + K_{m^2}}\, T_d$$

$$\Rightarrow \left(f_{eq}\omega_{m,ss} + T_d\right) + \frac{K_{m^2}}{R_a}\omega_{m,ss} = \frac{K_m}{R_a} e_a -----(7)$$

We now recall equation (3) and set D=0 for steady state to get.

---

[8] What is the significance and interpretation of the negative sign between the two boxes?
[9] In writing this equation, it is assumed that $K_m = K_b$. How is this justified?

$$T_{m,ss} = T_d + f_{eq}\,\omega_{m,ss} \;-\;-\;-\;-\;-(8)$$

$$\therefore\, T_{m,ss} = \frac{K_m}{R_a}\,e_a - \frac{K_m{}^2}{R_a}\,\omega_{m,ss} = K_1 e_a - K_2\omega_{m,ss} \;-\;-\;-\;-\;-(9)$$

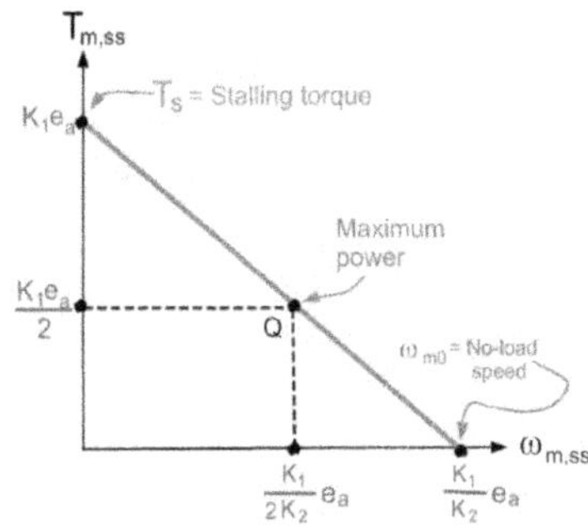

Equation (9) may be represented as a straight line for a given value of excitation voltage, e_a as shown in Figure beside.

Steady-State Power $P_{ss} = T_{m,ss}\,\omega_{m,ss}$

$$P_{ss} = K_1 e_a \omega_{m,ss} - K_2\omega_{m,ss}^2$$

For Maximum power, $\dfrac{dP_{ss}}{d\omega_{ss}} = 0$

$$\Rightarrow \omega_{m,ss} = \frac{K_1}{2K_2}e_a$$

This Speed is specified as "rated speed "of motor by the motor vendors. For maximum Power at point Q,

$$T_{m,ss} = \frac{T_s}{2} = \frac{K_1}{2}e_a$$

$$\omega_{m,ss} = \frac{\omega_{m0}}{2} = \frac{K_1}{2K_2}e_a$$

Example: ACDC Motor Subjected to Load Torque

An armature controlled DC-Servomotor, when supplied with 50v, armature voltage, develops a power of 2kw and runs at 1800 rpm. The equivalent inertia is 0.1kg-m², the equivalent viscous coefficient is 0.8 N-S-M/rad. Find the time constant of the motor, when the motor is running at this steady state of 1800 rpm and 50 volts, a torque of 2 N-m is applied. What is the new steady state speed?

<u>**Solution:**</u> Notice that this problem is very similar to the water heater problem in <u>Notes -06</u>. Just like that problem, one set of steady state data are given and new steady state is to be computed when a disturbance acts on the systems. Just see how many system parameters are computed: Motor constant, Resistance, K1, K2, stalling torque and no load speed.

<u>**Solution:**</u> Equation of the torque speed line is

$$T_{m,ss} = K_1 e_a - K_2 \omega_{m,ss} \quad ----(A)$$

Steady state speed = 1800 rpm   or $\omega_{m,ss} = \dfrac{2\pi * 1800}{60} = 60\pi \dfrac{rad}{sec}$

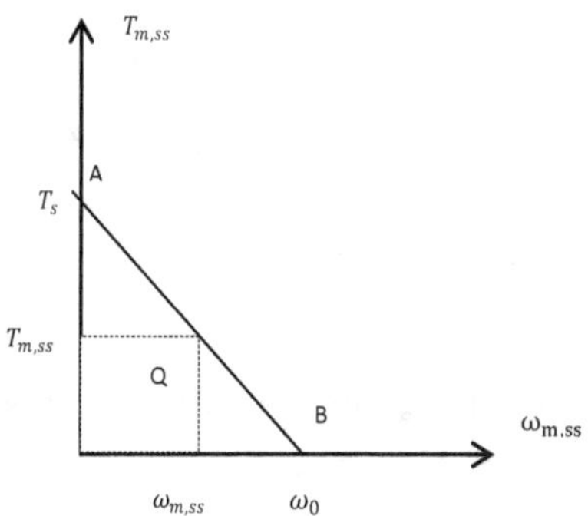

$P_{m,ss} = 2000 \ W = T_{m,ss} \omega_{m,ss} \Rightarrow T_{m,ss} = \dfrac{2000}{60\pi} = 10.6 \ N - M$

$\therefore T_s = 2T_{m,ss} = 21.2 \ N - m$ and $\omega_{m0} = 2\omega_{m,ss} = 120\pi \ rad/sec$

At point A, $21.2 = K_1 \times 50 - K_2(0) \Rightarrow K_1 = 0.424 = \dfrac{K_m}{R_a}$

At point B, $0 = (0.424 \times 50)K_1 - K_2(120\pi) \Rightarrow K_2 = 0.0526 = \dfrac{K_m^2}{R_a}$

We have We have $\dfrac{K_2}{K_1} = K_m = 0.13 \dfrac{N-M}{Amp}$. Thus $R_a = \dfrac{K_m}{K_1} = 0.31\Omega$

Given $J_{eq} = 0.1 \ kg - m^2$ and $f_{eq} = 0.8 \ N - s - m/rad$

We compute the time constant as: $\tau_0 = \dfrac{J_{eq} R_a}{f_{eq} R_a + K_m^2} = \dfrac{0.1*0.31}{(0.8*0.31)+0.13^2} = 0.12$

What are the units of time constant?

For solving the second part of the problem, note that speed is given by:

$$\omega = \dfrac{K_m/(f_{eq} R_a + K_m^2)}{1 + \tau_0 D} e_a - \dfrac{R_a/(f_{eq} R_a + K_m^2)}{1 + \tau_0 D} T_d$$

When a change occurs, ω changes. We are given that a change in T_d occurred and

$\Delta T_d = 2\,N - m$. When this change occurred, other input e_a remains same. So, $\Delta e_a = 0$. Taking difference operator on both sides,

$$\Delta\omega = \frac{K_m/(f_{eq}\,R_a + K_{m^2})}{1 + \tau_0 D}\,\Delta e_a - \frac{\dfrac{R_a}{f_{eq}\,R_a + K_{m^2}}}{1 + \tau_0 D}\,\Delta T_d = -\frac{2R_a}{(f_{eq}\,R_a + K_{m^2}}\,\frac{1}{1 + \tau_0 D}$$

At steady state all transients "die" SOD=0

$$\therefore \Delta\,\omega_{m,ss} = -\frac{2R_a}{f_{eq}\,R_a + K_{m^2}} = -2.34\,\frac{rad}{sec}$$

$$\therefore \text{New steady} - \text{state speed} = (60\pi - 2.34)\,\frac{rad}{sec}$$

Make sure that you understood the matter covered very clearly. At least make sure that you can easily reproduce the discussion without looking into the notes and you can solve the problem without looking at solution.

Example: ACDC Motor in Closed Loop

Schematic below shows an armature controlled DC–servomotor used in position control system. Draw mathematical block diagram and find transfer function relating θ_0 to θ_r .

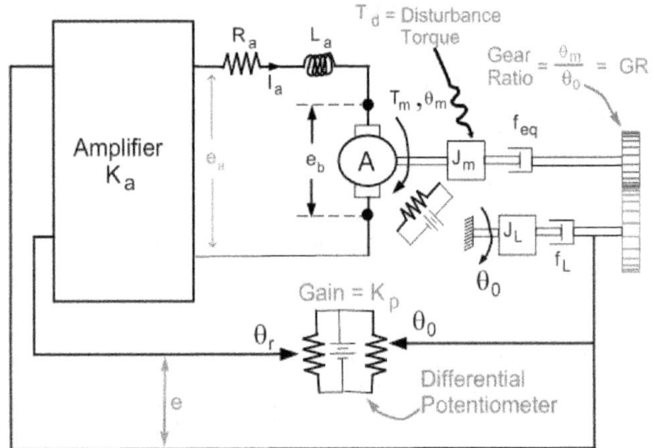

Solution: Start writing the equations beginning with the differential potentiometer and then write equations for Amplifier, armature circuit, motor action, load dynamics, and back emf equations. Then, "hook" up individual block diagrams to form one block diagram relating θ_0 or θ_r .

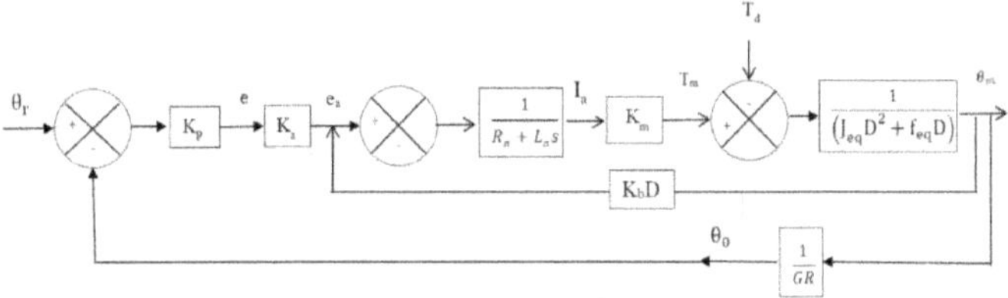

Home Work 09

(Q1). The parameters of an armature controlled DC- servomotor are as given below:

Equivalent inertia $= 0.5\ kg - m^2$

Equivalent viscous coefficient $= 0.1\ N\text{-}M\text{-}S\ /\ rad$

Motor constant $= 0.25\ \dfrac{N-M}{Amp}$

Motor armature resistance $= 0.5\Omega$

Motor efficiency $= 90\%$

If excitation voltage is 100v, what is the rated speed and rated power? What is the time constant of the motor?

When the motor is running at rated speed, a disturbance torque of 10 N-M is imposed on the motor. What will be the new study state speed?

(Q1). An armature controlled DC servomotor is connected to reduction rare box. The above combination can be assumed to be a device which provides an output torque directly proportional to armature voltage. Inertia and viscous damping are the only dynamic characteristics of the system. Tests show that starting with the motor stationary, the output shaft of the gear box reaches a speed of 3 RPM within $\dfrac{1}{18}th$ second when a step input of 50V is put to armature. The steady state speed is observed to be 6RPM. Derive an expression for TF of this system.

(Q2). The position of load is to be controlled using armature controlled DC servomotor, the performance specifications are:

a) The steady-state error to be within 2% of the input (step input)

b) Percentage overshot not to exceed 15%

c) The settling time is 1 second

Draw a schematic diagram of the system and corresponding mathematical block diagram. Describe the working of the system. Justify the selection of various parameters. The load is inertial and viscous type. Neglect inductance of armature coil.

1.10. Field Controlled DC-Servomotor

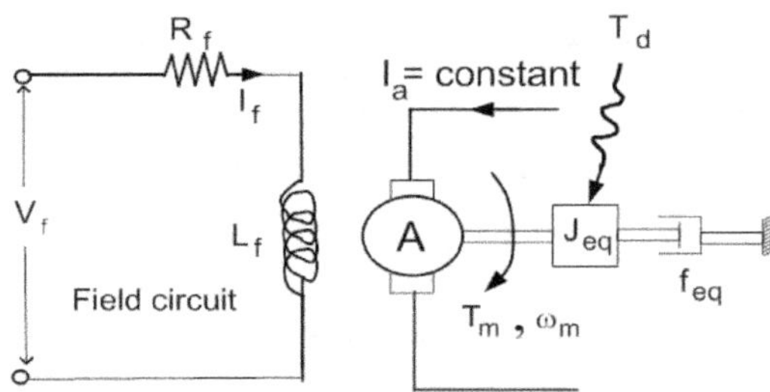

In the case of FCDC-servo motor, armature current is kept constant and required variations in motor torque are achieved by correspondingly varying the field voltage. Schematic of the FCDC-Servomotor may be shown in figure beside.

Notice that we have not shown the back emf in armature circuit. Since armature current has to be kept constant, armature voltage has to be adjusted in such a way that I_a= constant.

Considering different parts of the system viz., field circuit, motor effect, and load section, equations may be written as:

For field circuit: $V_f = (R_f + L_f D)I_f$

For motor effect: $T_m = K_{fc} I_f$

Load and inertia etc: $T_m = T_d + (J_{eq} D + f_{eq})\omega_m$

These equations can be represented in block diagram as:

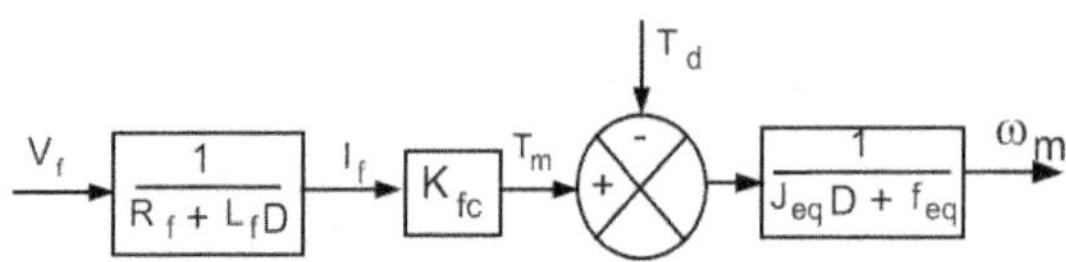

Notice, again, that the <u>feedback path present</u> in ACDC servomotor is absent here. Why?

$$\omega_m = \frac{K_{fc}}{(R_f + L_f D)(J_{eq} D + f_{eq})}V_f - \frac{1}{J_{eq} D + f_{eq}}T_d$$

If disturbance torque is set to zero T_d = 0, and field inductance is ignored, $L_f \approx 0$,

$$\omega_m = \frac{k_{fc}}{R_a J_{eq} D + R_a f_{eq}}V_f = \frac{(K_{fc}/R_a f_{eq})}{1 + \tau_f D} \quad \text{where} \quad \tau_f = \frac{J_{eq}}{f_{eq}}$$

Recall that the time constant of ACC servomotor is $\tau_a = \dfrac{J_{eq} R_a}{f_{eq} R_a + K_{m2}}$ for the same inertial and viscous load. We can at once see that

$$\tau_f = \frac{J_{eq}}{f_{eq}} = \frac{J_{eq}R_a}{f_{eq}R_a} > \frac{J_{eq}R_a}{f_{eq}R_a + K_{m2}} = \tau_a \quad or \quad \tau_f > \tau_a$$

That is ACDC servo motor settles <u>quicker</u> than FCDC servomotor. One reason we may specify is that in general, feedback action results in faster response and ACDC servomotor has <u>inherent feedback.</u>

Comparison of FCDC and ACDC Servomotors

1. It is difficult to maintain constant current source to keep I_a= constant whereas it is relatively easy to maintain I_f constant (why?)

2. Back emf acts as inherent damping in ACDC servomotor whereas back emf is "nullified" in keeping I_a= constant.

3. FCDC servomotor requires higher power source to keep I_a = constant and hence are generally driven directly from generators. Whereas ACDC servomotor can be driven through amplifiers.

4. Time constant FCDC servomotor is greater than that of ACDC servomotors i.e., FCDC takes more time to settle than ACDC.

FCDC servomotors can be used in controlling the position/velocity/acceleration of inertial/viscous loads much the same way as done with. The Homework problem illustrates this.

Home Work 10

(Q1). The following data refer to a position control system using field controlled DC-servomotor.

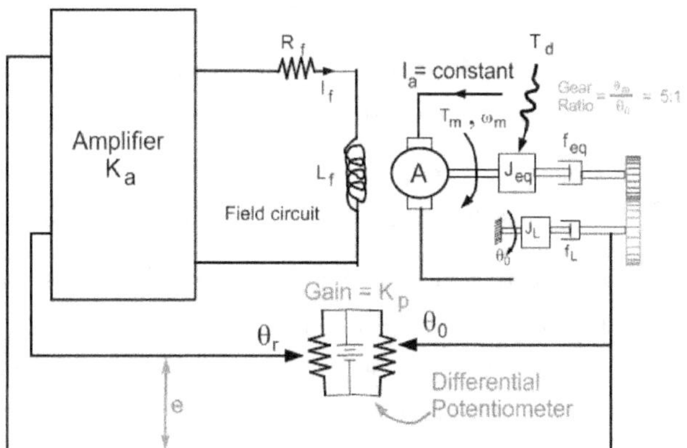

1. The error detecting device gives 1v per radian of error between its input and output.

2. Amplifier gain is 100 mA/v.

3. Servomotor output torque is 100N-m/Amp.

4. Inertia of motor rotor = $2.5*10^{-3}$ kg-m² and Inertia of load=$50*10^{-2}$ kg-m².

5. Speed reduction box between motor and load has a reduction ration of 5:1

6. Viscous function at load shaft is 1.0 N-M/(rad/sec).

Neglecting the inductance of field coil and viscous friction in the motor, find:

1. TF of closed loop system.

2. The magnitude of first overshoot when subjected to a unit step input.

Is it possible to solve this problem even when R_f and L_f are Not given?

1.11. AC – Servomotor

An AC – servomotor is basically a two phase induction motor with certain special design features. A two phase induction motor consists of two sector windings oriented at 90° in space and excited by a.c. voltages which differ in time phase by 90°:

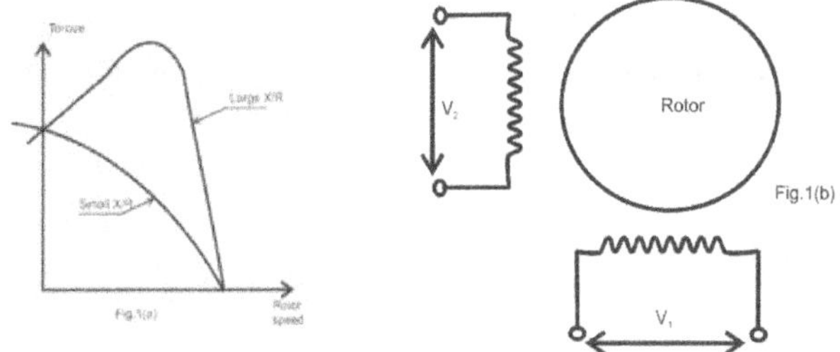

The shape of the torque speed characteristic depends on the ratio of rotor reactance to rotor resistance, i.e., $\frac{X}{R}$. In normal induction motor $\frac{X}{R}$ ratio is generally kept high so as to obtain maximum torque close to the operating region which is usually around 5% slip.

Mathematical Model and Block Diagram

A two – phase servomotor is different from a normal induction motor in two ways:

1) The rotor of the servomotor is built with high resistance so that $\frac{X}{R}$ ratio is small and torque-speed characteristics are ass in Fig. 1(a).

2) In servo applications, one of the phases, known as <u>REFERENCE PHASE</u> is excited by a constant voltage and the other one known as <u>CONTROL PHASE</u> has a variable magnitude and polarity supplied from a <u>servo amplifier</u>.

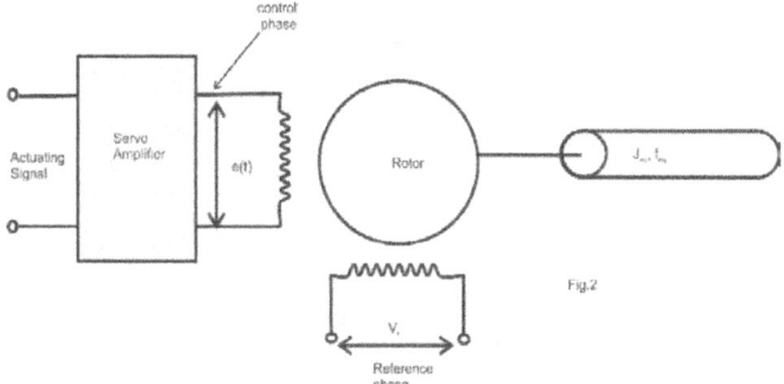

For the configuration shown in Fig. 2, the steady – state torque – speed characteristics are almost linear as shown in Fig. 3.

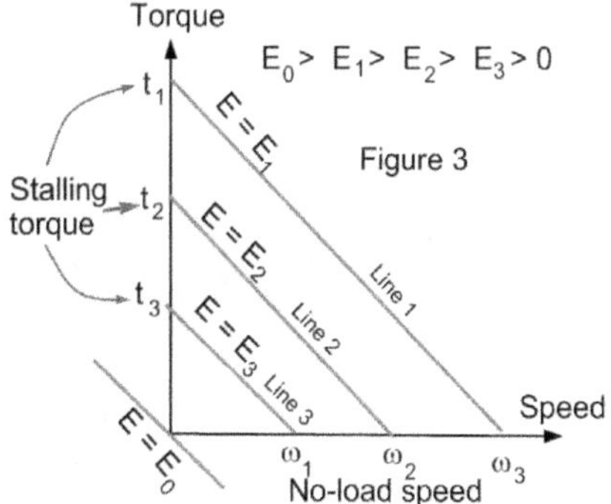

For each value of E, we have a no – load speed and a stalling torque associated.

The characteristics shown may be represented by the equation:

$$t_{m,ss} = K_1 E - K_2 \omega_{m,ss} \quad \text{---------------(1)}$$

Where E is the r.m.s. value of reference voltage in Figure 2, $t_{m,ss}$ is the steady-state torque and $\omega_{m,ss}$ is the steady-state speed.

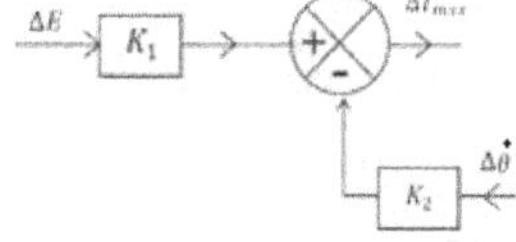

For drawing the block diagram of system shown in Fig. 2, we note, from equation (1).

$$\Delta t_{mss} = K_1 (\Delta E) - K_2 (\Delta \dot{\theta})$$

And the, we see that for load, we have,

$$t_{m,ss} = \left(J_{eq}D + f_{eq}\right)\dot{\theta} \;\Rightarrow\; \Delta t_{m,ss} = \left(J_{eq}s + f_{eq}\right)(\Delta\dot{\theta}) \text{ -----------------(2)}$$

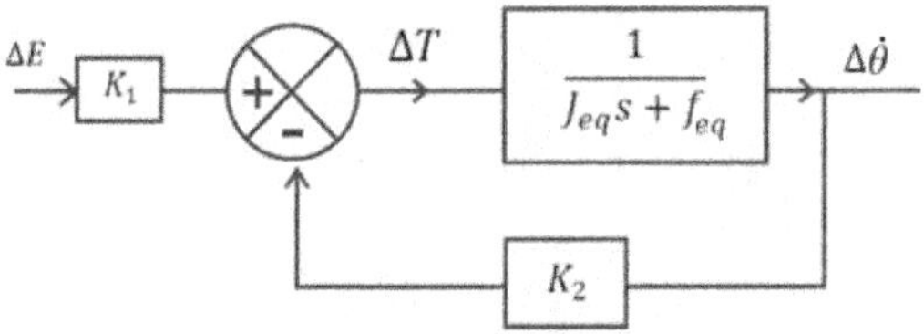

Simplifying equations (1) & (2), we get,

$$\frac{\Delta\dot{\theta}}{\Delta E} = \frac{K_1}{J_{eq}s + f_{eq} + K_2} = \frac{K_m}{1 + \tau_m s}$$

Where $K_m = \dfrac{K_1}{K_2 + f_{eq}}$ = motor gain constant

$\tau_m = \dfrac{J_{eq}}{K_2 + f_q}$ = motor time constant

The two block diagrams may be combined into one as shown in figure below:

Position Control Application

Having seen ACDC and FC DC servomotors in closed loop from for controlling the position of an inertial load, it is easy for us to draw the schematic of the position control system using AC – servomotor as:

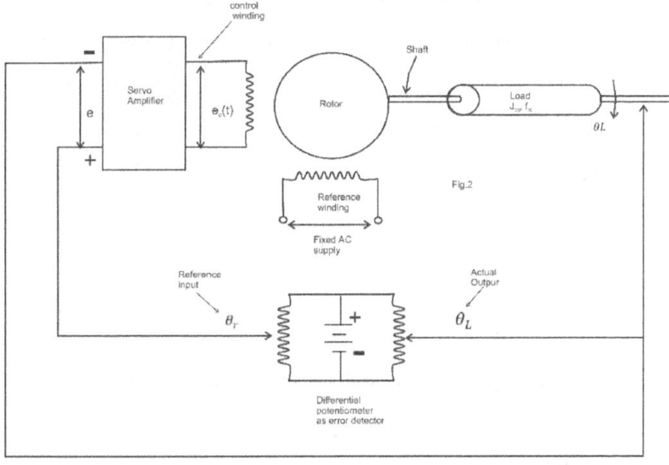

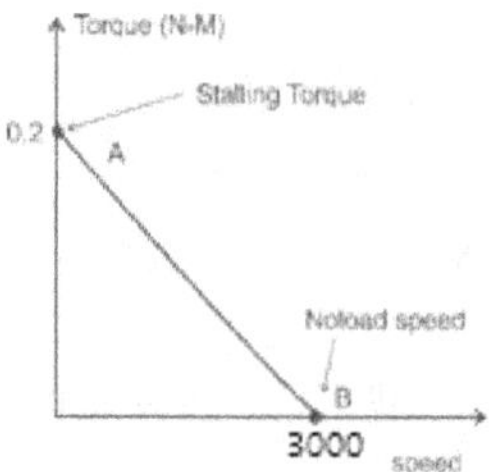

Example Problem

A two – phase servomotor has rated voltage applied to its reference winding. The torque speed characteristics of the motor with 115 volts, 50 Hz applied to the control winding exhibits linear torque speed of 3000 rpm. The moment of inertia of motor is 1 x 10⁻⁵ Kg-m² and friction is negligible. Find the TF relating shaft position θ to the control winding voltage ϑ_c.

Solution

At A, $0.2 = K_1(115) \Rightarrow K_1 = 0.00174$ (what are units?)

From equation $t = K_1(E) - K_2\omega$

At B, $0 = 0.2 - K_2\left(\frac{2\pi*3000}{60}\right)$

$\Rightarrow K_2 = 0.000636$ (units?)

$\therefore K_m = \frac{K_1}{K_2+f} = \frac{0.00174}{0.000636} = 2.73$ (units?)

$\tau_m = \frac{J}{0.000636} = \frac{1*10^{-3}}{0.000636} = 0.0157$ (units?)

$\therefore \frac{W}{E} = \frac{2.73}{1+0.0157\,s}$ and so, $\frac{\theta}{E} = \frac{2.73}{s(1+0.0157\,s)}$

Home Work – 11

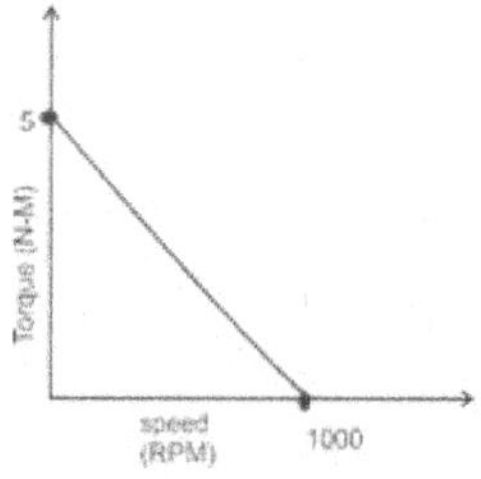

1) The torque speed characteristics of a two – phase AC servomotor is shown in Fig. beside. The voltage to fixed phase windings is 115 volts. If the inertia od rotating parts in the servomotor is 5 x 10⁻³ Kg-m², what value of damping coefficient is required so that the time constant of the speed control system is 1 second.

2) An AC servomotor has linear torque speed characteristics. The stalling torque is 10N-M and no load speed is 2000 rpm. The corresponding control winding voltage is 120 volts, 50 Hz. Such motor is used to control the position of a heavy inertial and viscous load through speed reduction gear box of 5:1 reduction ratio. The position transducer mounted on load shaft has a gain of 1 V per radian. The error signal is amplified before feeding to the control windings. What value of amplifier gain will you select to get to get the percentage overshoot to 30%. The load inertia is 1 kg -m² and damping coefficient is 0.02 kgf-m-s/rad. Draw the block diagram of the system.

3) An electric motor provides torque proportional to applied voltage. The motor drives load through speed reduction gear box of ratio 6:1. The load is inertial and viscous type. When a step voltage of 50 volts is applied, the load reaches a speed of 2 rad/sec within 0.5 seconds. The steady state speed of the load is 3 rad/sec. Find the transfer function $\theta_m(s)/V(s)$ i.e., the motor angular deflection to voltage applied. Neglect inertia of motor shaft. If torque produced by motor per volt is 0.01 kgf-m, find the inertial load and coefficient of viscous damping.

1.12. Operational Amplifiers

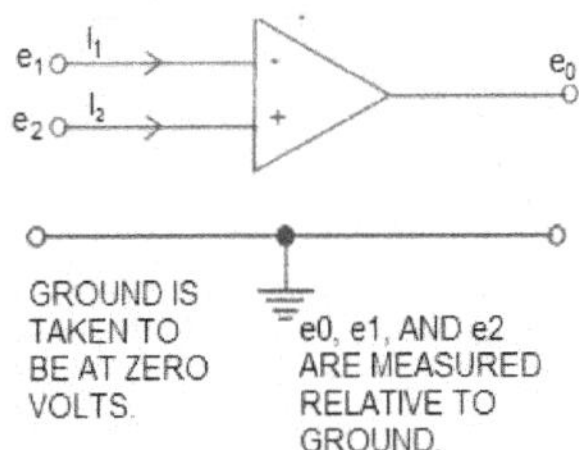

Operational amplifiers, abbreviated as OP amps, are frequently used to amplify signals in sensor circuits and also to implement compensators and filters.

There are two terminals, positive and negative, marked + and -. Input to – terminal is inverted in sign while that to the + terminal is not inverted. Thus the total input to the amplifier shown above is $(e_2\text{-}e_1)$.

Thus, for the amplifier, $e_0 = K(e_2 - e_1)$. Also, it may be noted that the amplifier doesn't draw any currents into its terminals, thus $i_1 = 0, i_2 = 0$.

The constant K is called the _differential voltage gain_ of the amplifier. The inputs e_1 and e_2 may be positive or negative and may be dc or ac signals. K is very high, of the order of 10^5 to 10^6 for dc signals. For ac signals, K is equally high for frequencies up to almost 10kHz. The differential gain K decreases as signal frequency increases and almost becomes unity for frequencies of the orders 1 MHz to 50 MHz. by choosing the grounding point and inserting electrical elements (resistors and capacitances), the operational amplifiers may be used to implement filters and compensators.

A few important observations regarding op amps are apt here before we see a few applications:

- No current flows into the input terminals.
- The output voltage is not affected by the load connected to the output terminal.
- The input impedance is infinity and output impedance is Zero for ideal op amp.
- In an actual op amp, very small currents flow into the terminal.
- Actual op amps shall NOT be connected to too much of load.
- Op – amps are active elements: they need external power.
- In all examples we see, we will assume that the op amps we deal with are ideal op amps.

Since the gain of the amplifier is too high, it is necessary to have negative feed back to make it stable.

Food for thought: Why is high gain a cause of instability?

What is meant by high gain in physical terms?

Is there a connection between high gain and high sensitivity? At least in physical context?

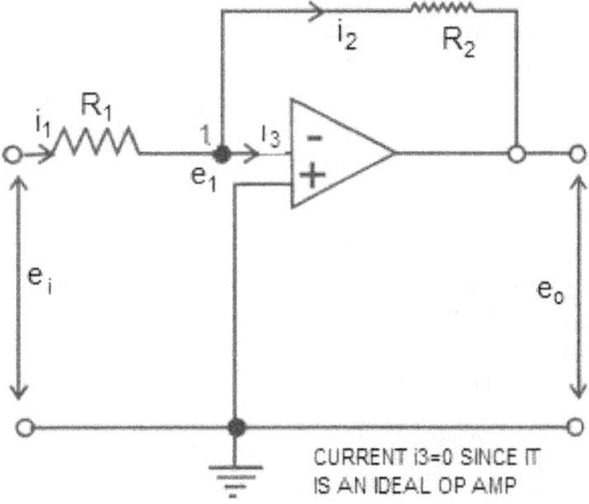

We will see a few examples of operational amplifiers to implement inverting, non – inverting amplifiers, filters and compensators.

Example 1: Inverting Amplifier

Find the TF of the op amp shown in Figure.

Solution: Firstly note what are the inputs to the positive and negative terminals of op amp. The positive terminal is connected to ground so its input voltage is Zero. The negative terminal has an input voltage of e_1. Therefore, if K is the differential gain of the amplifier, we have

Amplifier eqn: $e_0 = K(0 - e_1) \Rightarrow e_1 = -\dfrac{e_0}{K}$

At node 1, we see that $i_1 = i_2$ since $i_3 = 0$. [why $i_3 = 0$?]

Therefore

Current eqn: $\dfrac{e_i - e_1}{R_1} = \dfrac{e_1 - e_0}{R_2} \Rightarrow \dfrac{e_i + \frac{e_0}{K}}{R_1} = \dfrac{\frac{-e_0}{K} - e_0}{R_2}$

$$\Rightarrow e_i = -e_0 \left(1 + \frac{1}{K}\right)\frac{R_1}{R_2}$$

For ideal op amp, $K \to \infty$. Therefore, $e_0 = -\dfrac{R_1}{R_2} e_i$. Upon taking Laplace Transform, we see that

$$G(s) = \frac{E_0(s)}{E_i(s)} = -\frac{R_2}{R_1}$$

Since the sign of input is negated, it is called *inverting amplifier*.

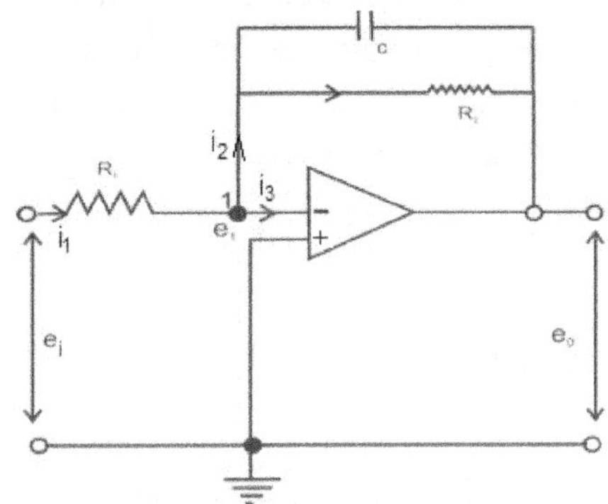

+VE TERMINAL IS CONNECTED TO GROUND SO ZERO VOLTAE. -VE TERMINAL IS CONNECTED TO VOLTAGE e1.

Example 2: First Order Filter

Obtain the TF $\dfrac{E_0(s)}{E_i(s)}$ for the electrical circuit shown beside.

Solution:- Let z be the net impedance of R2 and C. then, we have

$$\frac{1}{z} = \frac{1}{R_2} + Cs$$

The amplifier equation and the current equation may be written as:

Amplifier equation: $e_0 = K(0 - e_1) \Rightarrow e_1 = -\frac{e_0}{K}$ .... (1)

Current equation: $\frac{e_i - e_1}{R_1} = \frac{e_1 - e_0}{Z} = (e_1 - e_0)\frac{1+R_2Cs}{R_2}$

Using equation (1), $\frac{e_i + \frac{e_0}{K}}{R_1} = -(\frac{e_0}{K} + e_0)\frac{1+R_2Cs}{R_2}$

$$\Rightarrow e_0\left[\frac{1}{K} + \left\{\frac{1}{K} + 1\right\}\frac{1+R_2Cs}{R_2}R_1\right] = -e_i$$

For an ideal amplifier, $K \to \infty$

$$\therefore G(s) = \frac{E_0(s)}{E_i(s)} = -\frac{R_2}{R_1}\frac{1}{(1+R_2Cs)}$$

This circuit inverts the sign and implements a First order filter. If such sign inversion is NOT desired, and inverting amplifier in example 1 may be connected to the output.

Example 3: PID Controller

Find the transfer function $\frac{E_0(s)}{E_i(s)}$ for the electrical circuit shown below:

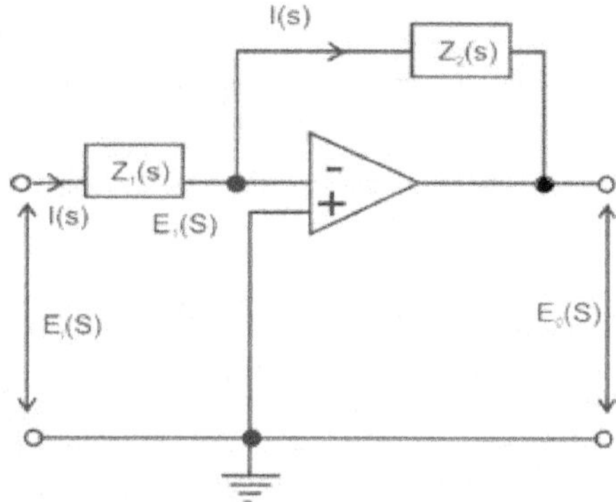

Solution

Amplifier equation: $E_o(s) = K(0 - E_1(s)) \Rightarrow E_1(s) = -\frac{E_0(s)}{K}$

Current equation: $\frac{E_i(s) - E_1(s)}{Z_1(s)} = \frac{E_1(s) - E_0(s)}{Z_2(s)}$

$$\Rightarrow \frac{E_i(s) + \frac{E_0(s)}{K}}{Z_1(s)} = \frac{-\frac{E_0(s)}{K} - E_0(s)}{Z_2(s)}$$

$$\Rightarrow E_1(s) = -E_0(s)\left[\frac{1}{K} + \left\{\frac{1}{K} + 1\right\}\frac{Z_1(s)}{Z_2(s)}\right]$$

As $K \to \infty$, $G(s) = \frac{E_0(s)}{E_i(s)} = -\frac{Z_2(s)}{Z_1(s)}$

Notice that Example 1 and Example 2 are special cases of Example 3. In example 1, $Z_1(s) = R_1$ and $Z_2(s) = R_2$. In example 2, $Z_1(s) = R_1$ and $Z_2(s) = \frac{R_2}{1+R_2Cs}$. It is possible to construct Proportional, Proportional + Derivative, Proportional + Integral, Proportional + Derivative + Integral, Lead, lag, and lag-lead control actions using the Example 1, 2, and 3 by using Resistances and capacitance in appropriate manner.]

Home Work 12

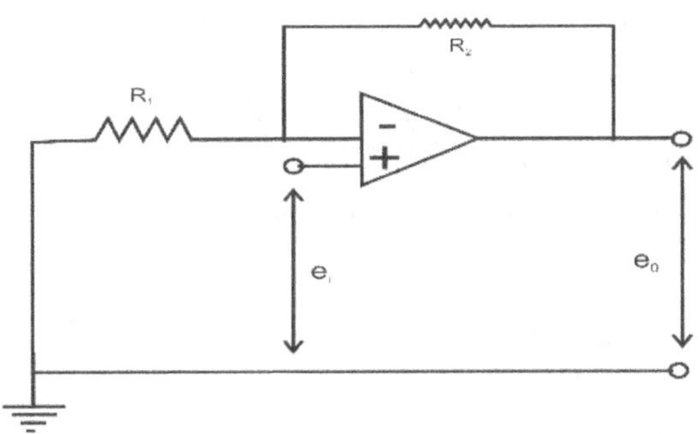

1) Show that the electrical circuit shown below can be used as non – inverting amplifier. Find the gain e_0/e_i of this circuit. Assume that the op – amp is ideal

2) Show that the circuit shown below can be used as lead or lag network and show that its TF is obtained as:

$$\frac{E_0(s)}{E_i(s)} = K_c \frac{s+\frac{1}{T}}{s+\frac{1}{\alpha T}} \quad \text{where } T = R_1C_1, \quad \alpha T = R_2C_2, \text{ and } K_c = \frac{R_4C_1}{R_3C_2}.$$

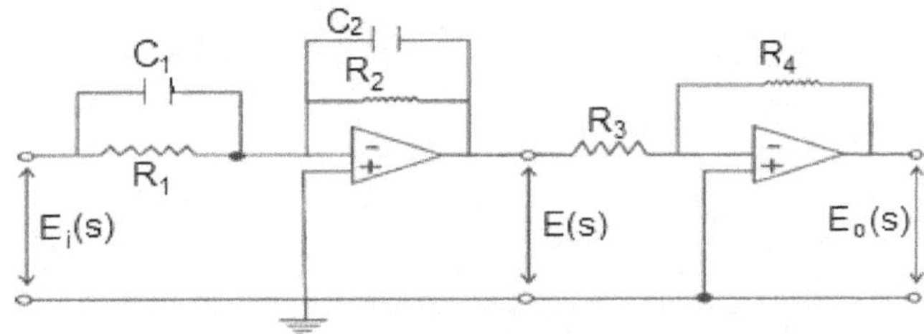

3) Show that the circuit shown below is an electrical PID controller with $K_p = \frac{R_4(R_1C_1+R_2C_2)}{R_3R_1C_2}$ = proportional gain, $T_i = \frac{1}{(R_1C_1+R_2C_2)}$ = integral time,

$T_d = \frac{R_1C_1R_2C_2}{(R_1C_1+R_2C_2)}$ = derivative time, and $\frac{E_0(s)}{E_i(s)} = K_p\left[1 + \frac{T_i}{s} + T_d s\right]$

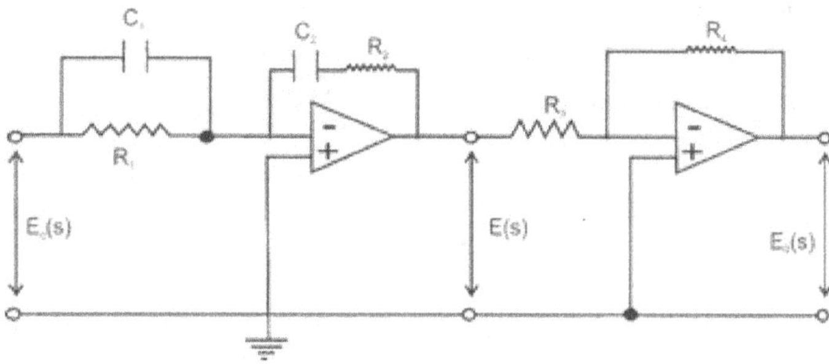

1.13. Block Diagrams

While discussing dynamic models of open loop and closed loop systems, we have seen that the equations describing systems can be put in a BLOCK DIAGRAM FORM to obtain pictorial representation. We will formalize the discussion of block diagrams and present the rules of BLOCK DIAGRAM REDUCTION. These rules are used to obtain overall transfer function of a system represented by a block diagram. First, we present notation.

Notation

1. *Line Segment*

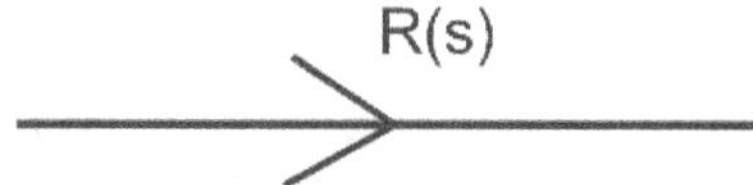

Directed line segment in a block diagram indicates a variable, sometime also called a SIGNAL. For example, to represent a signal (or a variable R(s)) we draw a directed line segment as shown beside.

2. *TF Box or Gain*

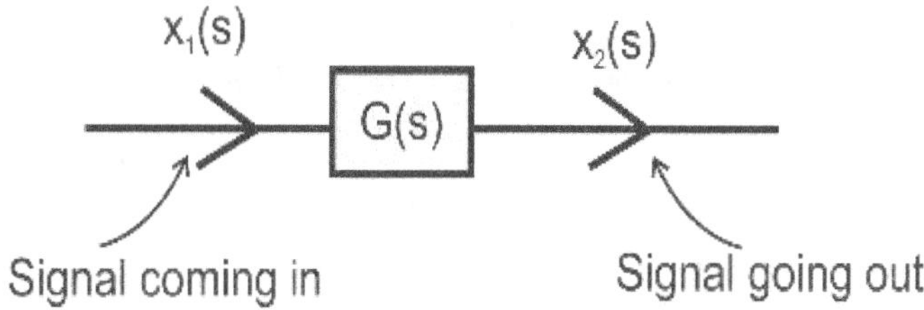

T.F. Box indicated by a rectangular block indicates multiplication. The variable (or signal) coming in is multiplied by TF in the box and gives the signal going out. For example, the diagram shown beside indicates $X_1(s)G(s) = X_2(s)$

3. *Summing Point*

When there is a need to add or subtract two signals, we use summing point. When subtraction is involved, it is also referred to as COMPARATOR.

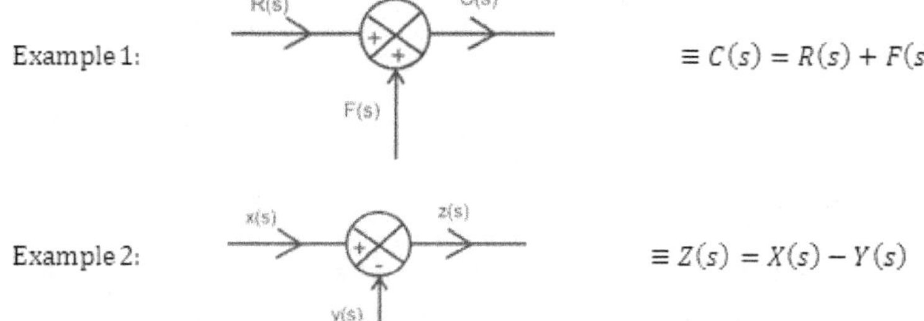

Example 1: $\equiv C(s) = R(s) + F(s)$

Example 2: $\equiv Z(s) = X(s) - Y(s)$

NOTICE THE POSITIVE AND NEGATIVE SIGNS IN THE BLOCK DIAGRAM AND HOW THEY ARE REFLECTED IN EQUATIONS.

4. *Branch Point*

Branch point is used in situations where the same variables (or signal) needs to be used at more than one place of block diagram.

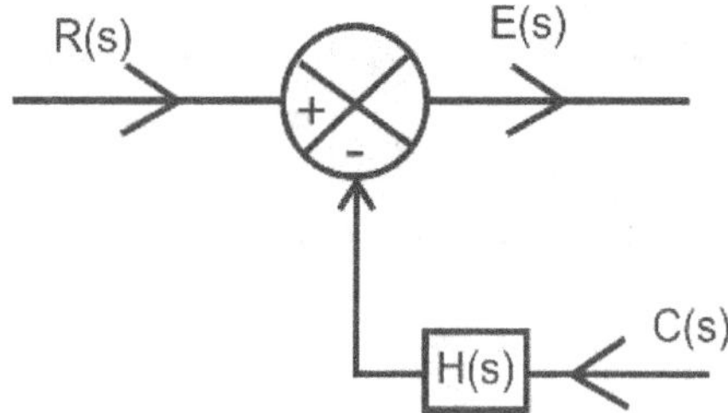

We will see one example of a system od equations and represent them in a block diagram. Consider the following equations involving input R(s), output, C(s), and Transfer Function G(s) and H(s).

$$E(s) = R(s) - H(s)C(s)$$

$$C(s) = G(s)E(s)$$

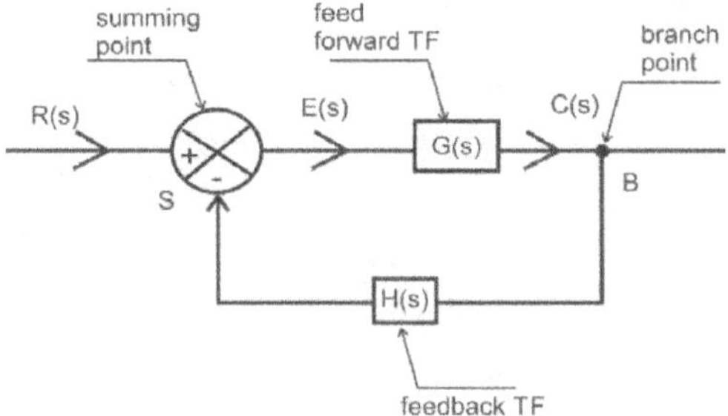

The two equations can be put into one block diagram shown besides.

The block diagram shown here has one summing point, one branch point, one feed forward TF, one feed back TF, and one LOOP. But in general, block diagrams of complex systems may involve more of these. In such situations we use RULES OF BLOCK DIAGRAM REDUCTION to successively simplify portion of block diagram to obtain overall TF. This reduction is typically achieved in "stages" in each stage, certain rules are applied to ensure that the underlying equations are not affected. These are called Rules of Block Diagram Reduction.

Rules of Block Diagram Reduction

There are <u>eight rules</u> and <u>Two critical rules</u>, totaling to <u>TEN</u>. We will see them one by one.

1. *Rules 1: Blocks in series or CASCADE need to be multiplied*

2. *Rules 2: Blocks in parallel need to be "summed up" sign-wise.*

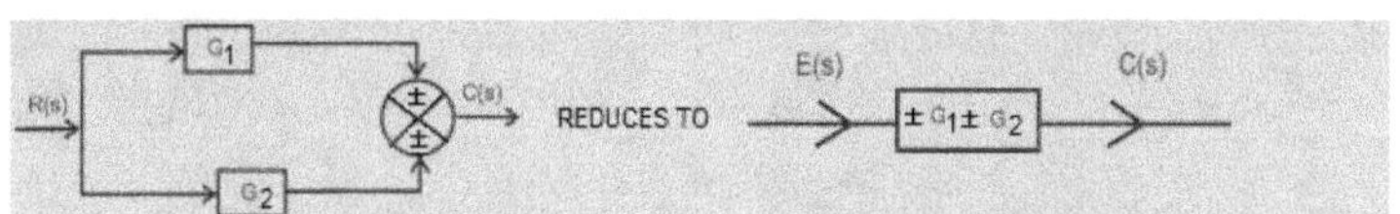

3. *Rule 3: Feedback loop may be eliminated using rule indicated.*

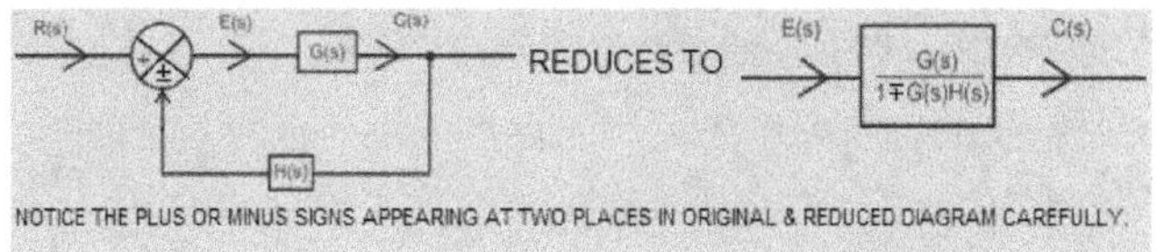

Once again, carefully note the signs in the summing point and the sign in the denominator of the transfer function and convince yourself thoroughly about these signs.

4. Rule 4: Associative Law for Summing

Locations of two consecutive summing points can be swapped.

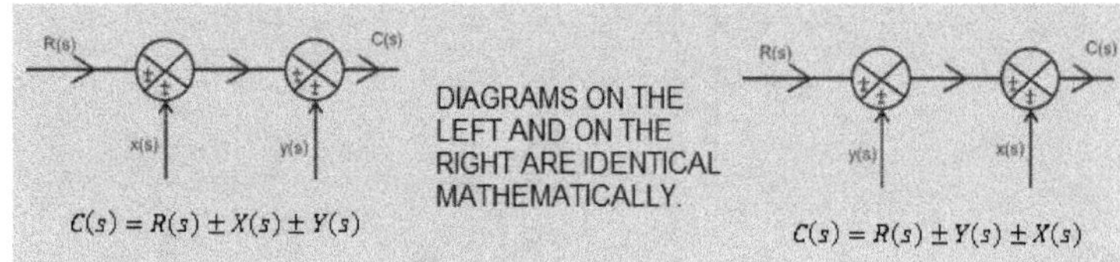

5. Rule 5: Moving of summing point before a block

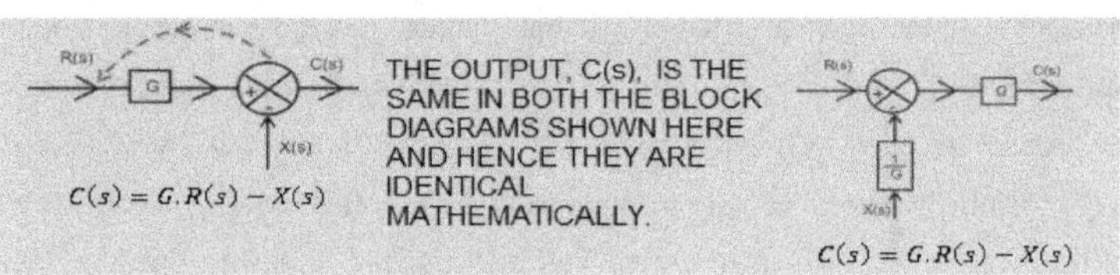

6. Rule 6: Moving summing point after a block

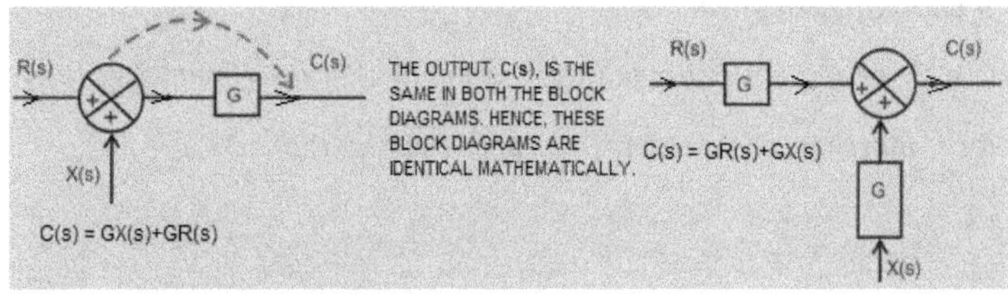

7. Rule 7: Moving branch take off point before a block

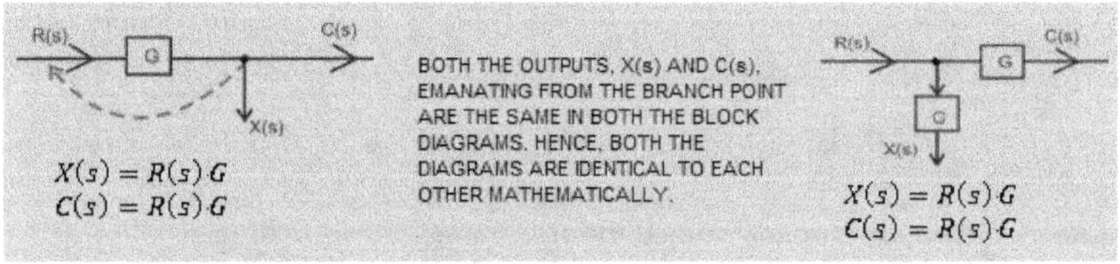

8. Rule 8: Moving a branch take off point after a block

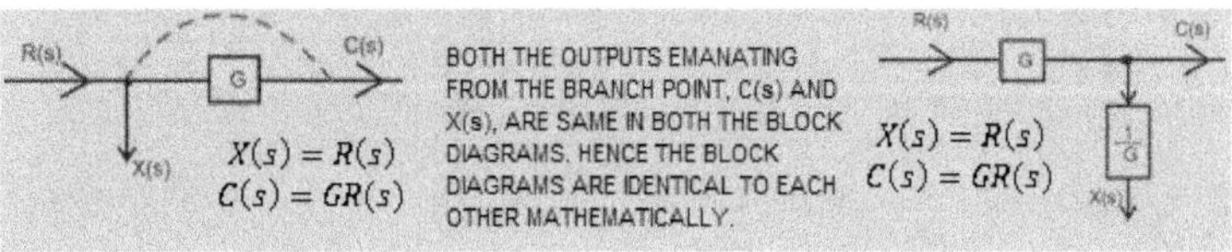

The next two are critical rules, rules where many students (and teachers) may tend to commit mistakes. Pay particular attention.

9. Rules 9: Shifting a take off point after a summing point

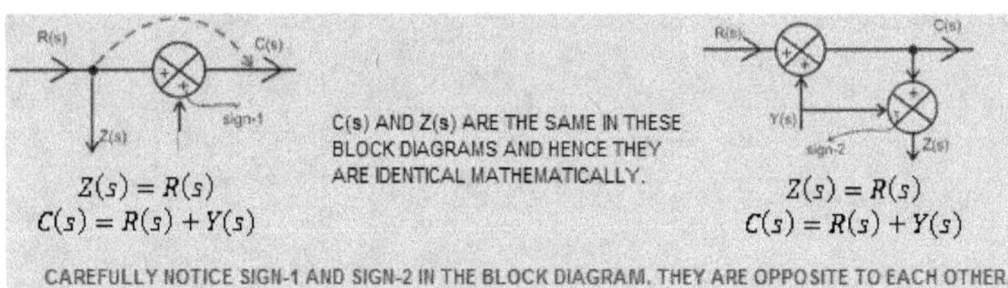

$$Z(s) = R(s)$$
$$C(s) = R(s) + Y(s)$$

$$Z(s) = R(s)$$
$$C(s) = R(s) + Y(s)$$

CAREFULLY NOTICE SIGN-1 AND SIGN-2 IN THE BLOCK DIAGRAM. THEY ARE OPPOSITE TO EACH OTHER.

10. Rules 10: Shifting a take off point before a summing point

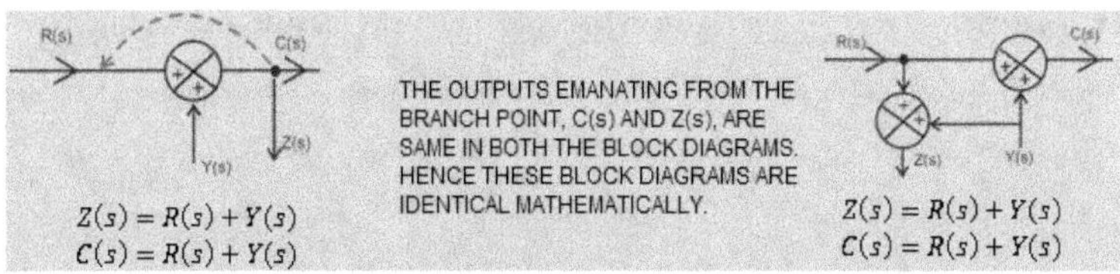

$$Z(s) = R(s) + Y(s)$$
$$C(s) = R(s) + Y(s)$$

$$Z(s) = R(s) + Y(s)$$
$$C(s) = R(s) + Y(s)$$

Each time we make a change to the block diagram, we need to ensure two things:

1. The block diagram is simplified in someway
2. The under lying equations are NOT affected.
3. That is, all the 10 rules specified above are satisfied.

It takes about ONE to TWO hours of practice to get good at this!

Example 1: Loop within a Loop

Simplify the block diagram shown below

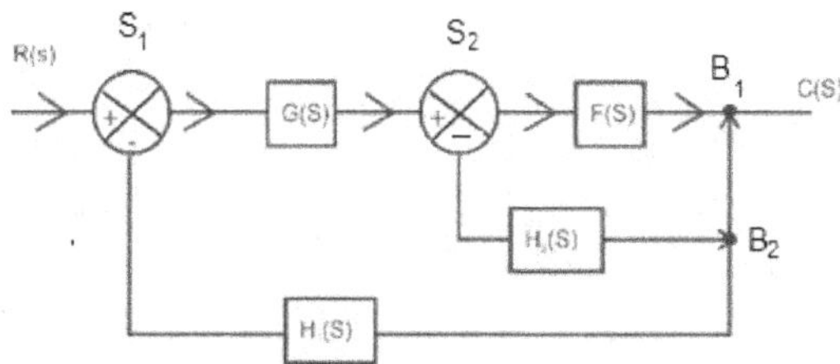

Solution

<u>Step 1</u>: Branch point can jump over a branch point without any hassle! Hence move B2 to a point just before B1 so that the block diagram simplifies to the following.

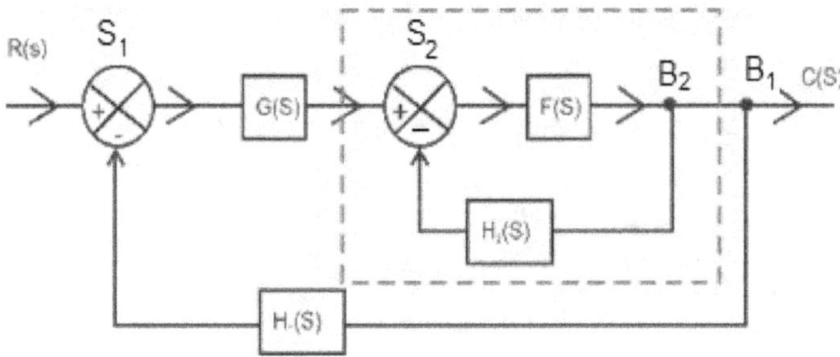

Notice the dashed box drawn over a segment of the block diagram. This is clearly seen to be a feedback loop specified in Rule 3. This can be simplified using Rule 3 to a single block with transfer function: $\frac{F(s)}{1+F(s)H_2(s)}$. With this simplification, the block diagram simplifies to the one shown in Figure below. After the completion of this step of simplification, the Summing point S2 and the Branch point B2 disappear from the picture.

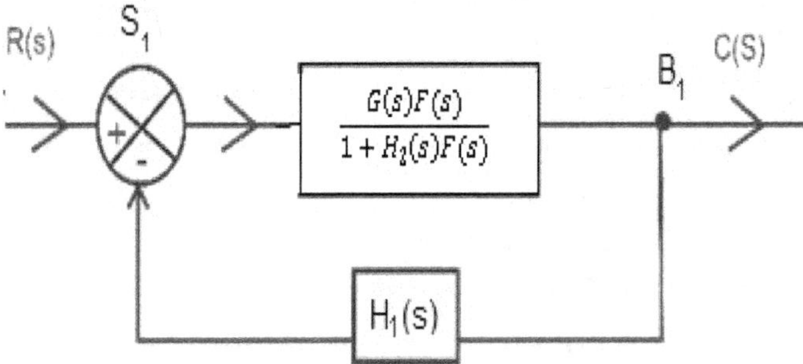

With step 1 of simplification, there will be two blocks with transfer functions: G(s) and $\frac{F(s)}{1+F(s)H_2(s)}$. These blocks are in series and hence can be multiplied and put in to one block using Rule 1 to yield: $\frac{G(s)F(s)}{1+H_2(s)F(s)}$.

Step 2: The block diagram obtained after step 1 is seen to be a feedback loop given in Rule 3. Hence the transfer function may be obtained as

$$\therefore \frac{C(s)}{R(s)} = \frac{\dfrac{F(s)G(s)}{1+H_2(s)F(s)}}{1+\dfrac{F(s)G(s)H_1(s)}{1+H_2(s)F(s)}} = \frac{F(s)G(s)}{1+H_2(s)F(s)+F(s)G(s)H_1(s)} .$$

Example 2:- Overlapping Loops

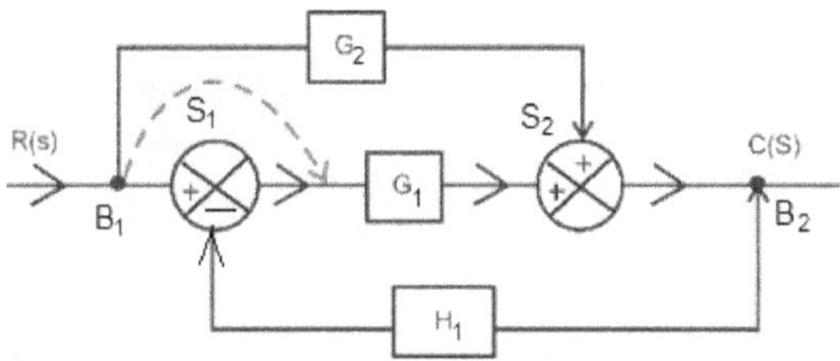

STEP-1: Move B_1 to the right of S_1. There is one extra summing point. Why did this have to come in? Also, the path from B1 to G2 to S2 is drawn from below. Does it matter?

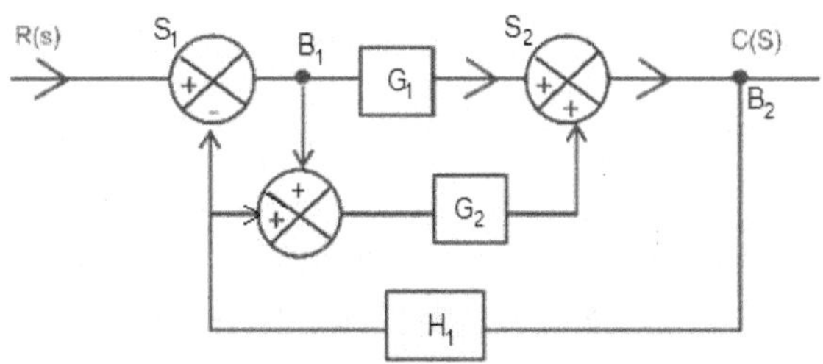

STEP-2: Rearrange. And notice that G_1 and G_2 and in parallel and hence can be summed.

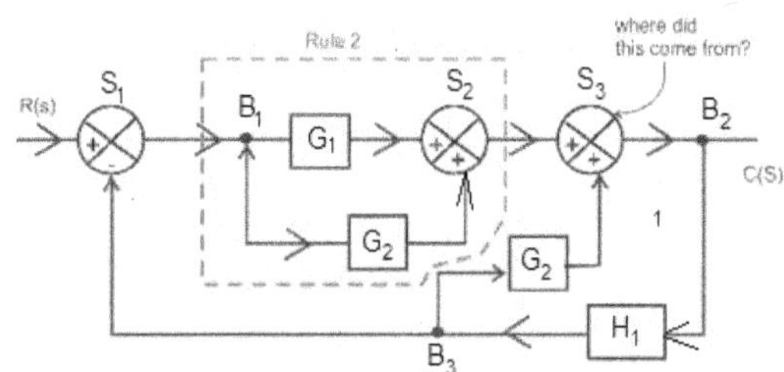

STEP-3: After completion of step 2, the block diagram appears as shown below..

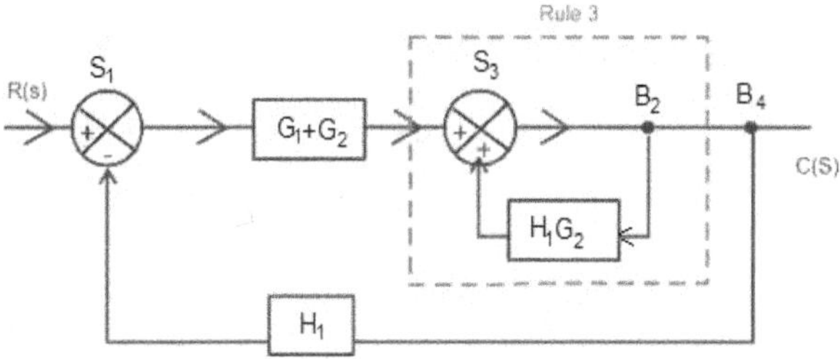

<u>Step 4</u>: Take care of loop S_3-B_2-S_3. With this, the block diagram appears as shown below.

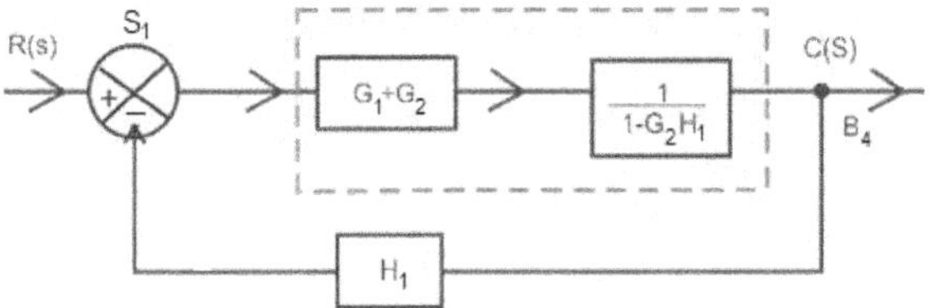

<u>STEP-5</u>:- First apply rule 1 to cascaded blocks and then eliminate feedback loop to get TF as

$$\frac{C(s)}{R(s)} = \frac{\dfrac{G_1 + G_2}{1 - G_2 H_1}}{1 + \dfrac{(G_1 + G_2)H_1}{1 - G_2 H_1}}$$

$$\text{Or,} \quad \frac{C(s)}{R(s)} = \frac{G_1 + G_2}{1 + G_2 H_1}.$$

To sum it up, block diagram reduction method may find its application in reducing complex block diagrams and thus find TF. This method uses pictorial reduction algebra given by TEN rules. Though it offers pictorial means of simplification of equations, as seen in example 2, this may involve drawing successive simplified block diagrams even in relatively simple cases such as example 2.

Signal flow diagram, discussed next, is a much more efficient way of finding TF from a set of equations. Mason's gain formula gives a one – go method for finding overall given or TF.

Home Work 13

Simplify the following block diagrams and hence find the Transfer function C(s)/R(s)

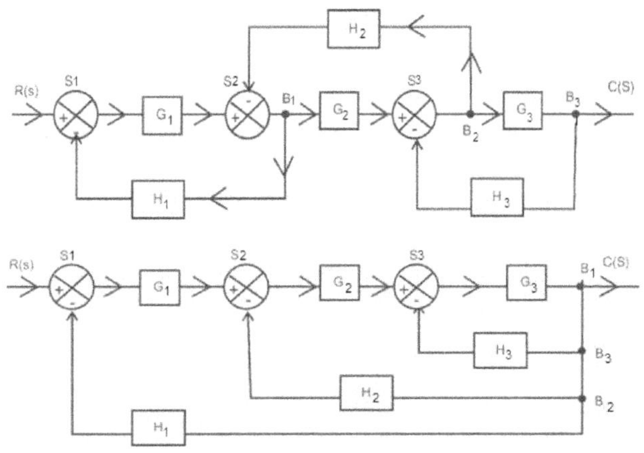

1.14. Signal Flow Graphs

Introduction

A Signal Flow Graph (SFG) is a *pictorial representation* of a *system of equations* describing the cause – and effect relations of a linear system modeled by algebraic equations. SFG shows variables (also termed as signals), gains (which are either Transfer functions or constant gains), and relations between variables in a compact form. SFG was introduced by S.J. Mason during 1950's and is a very effective tool for calculating the gain (or overall TF between an output variable and an input variable).

In an SFG, each variable is represented as a node, shown by a small circle (o), and each gain by a branch, show by a (straight or curved) directed line segment going from a node to another node. The "gain" between these two nodes is shown next to the branch. In an SFG, each node acts as a "summing point" and can *add* signals coming *into* it. As an example consider the equation: $y = \underbrace{G_1(s)x_1}_{\text{Term 1}} + \underbrace{G_2(s)x_2}_{\text{Term 2}}$.

In this equation, there are <u>three</u> variables (or signals) of which two variables (x_1 *and* x_2) are cause variables and one effect variable, y_1.. Since there are three variables, there will be three nodes in SFG shown below:

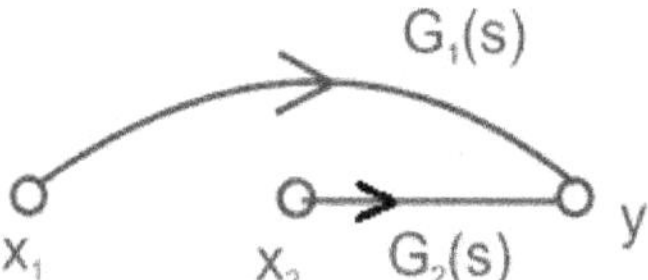

Since there are two terms in the equation, there will be two branches to represent the equation. Each branch goes from a cause variable to the effect variable: one branch goes from x_1 to y and the other branch goes from x2 to y.

<u>Example 2</u>:- Draw SFG to represent the set of equations shown below

$$E(s) = R(s) - H(s)C(s) \quad \begin{cases} \text{without loss of generality} \\ \text{we may write: } E(s) = 1{\cdot}R(s) - H(s){\cdot}C(s) \end{cases}$$

$$C(s) = G(s)E(s)$$

In this set of equations, R(s), E(s), and C(s) are variables[10] and - H(s) and G(s) are gains. We draw SFG in two stages for the sake of illustrator: one stage for one equation. In actual practice

---

[10] How can we decide in this case?

the final SFG can be shown incrementally. Carefully note the process of drawing the signal flow graph shown in figure below.

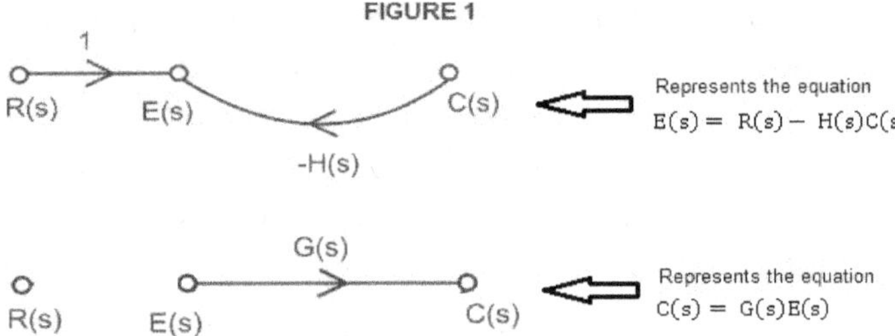

The above two graphs are combined into one graph shown below.

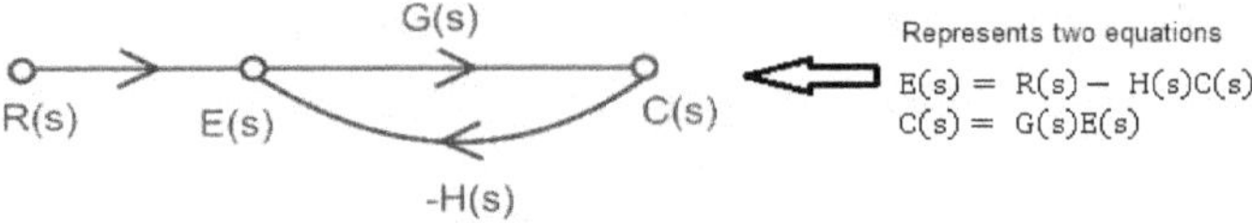

Notice that the branches come towards the effect-variable. For example in the first equation, E(s) is the effect variable and is shown on the left hand side of the equation. Correspondingly all branches go towards E(s). The node for E(s) sums up the two branches coming towards it to give R(s) - H(s)C(s).   Similarly in second equation, C(s) is the effect-variable and E(s) is the cause-variable. The branch goes from E(s) to C(s). The two graphs are combined into one graph as shown in the bottom-most picture.

We will sum – up what we have seen so far and then go for further discussion:

- SFG apples to linear systems.
- SFG deals with algebraic equations.
- Nodes represent variables.
- Branches represent gain.
- Signals Travel along branches **ONLY IN DIRECTION SHOWN.**
- Branch going <u>from node x to node y</u> represents dependence of y on x <u>but NOT THE REVERSE.</u>
- Signal travelling from node x to node y is multiplied by the gain of the branch and is <u>"collected"</u> by node y.
- The value of variable represented by a node is the <u>sum</u> of all signals <u>coming into it:</u>

We present a few definitions before we present <u>Mason's gain formula (MGF)</u> .

Definitions

1. INPUT NODE

Input Node is a node that has only out going branches. For example R(s) in fig. 1 is an input node or SOURCE.

2. OUTPUT NODE

Output Node is a node that has only incoming branches. Strictly speaking, there is no output node in Figure 1. But we can make an output node by adding an equation. For example, if we add the equation: $C_1(s) = 1 \cdot C(s)$, the SFG may be drawn as shown in Figure 2. Now, $C_1(s)$ is output node or SINK

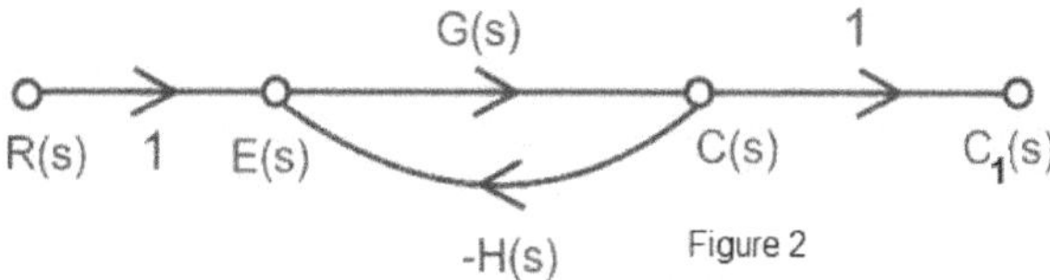

Figure 2

3. PATH

Path is any collection of a continuous succession of branches traversed in the same direction. R(s) - E(s) - C(s) is a path in above SFG. Similarly R(s) - E(s) - C(s) - E(s) and R(s) - E(s) – C(s) – $C_1(s)$ are also paths in SFG shown in Figure 2.

4. FORWARD PATH

Forward path is a path that starts at an input node and ends in an output node *without touching any intermediate node more than once*. In the SFG shown in Figure 2, there is only one forward path: R(s) - E(s) - C(s) – $C_1(s)$. Though the SFG in Figure 2 has only one forward path, there can be multiple forward paths in an SFG for more complicated systems.

5. FORWARD PATH GAIN

Forward Path Gain is the product of all gains of the forward path. For example, in the SFG shown in Figure 2, the forward path gain of path R(s) - E(s) - C(s) – $C_1(s)$ is G(s).

6. LOOP

Loop is a path that originates and terminates on the same node *without traversing any intermediate node more than once*. In SFG in Figure 2, there is only one loop: E(s)-C(s)-E(s)

7. *NON – TOUCHING LOOPS*

Two loops are said to be non – touching if there is no common node.

8. *LOOP GAIN*

Loop Gain is the product of gains encountered while traversing along a loop. For example, Loop gain of E(s)-C(s)-E(s) in Figure 2 is -G(s) H(s).

Mason's Gain Formula (MGF)

Mason's gain formula specifies the overall gain from an input node y_{in} to y_{out} as

$$M = \frac{y_{out}}{y_{in}} = \sum_{i=1}^{N} \frac{F_i \Delta_i}{\Delta}$$

Where

    M = gain between y_{in} and y_{out} .

    y_{in} = input node

    y_{out} = output node

    N = Total number of forward paths between y_{in} and y_{out}.

    F_i = gain of ith – forward path

    Δ = determinant of SFG

    = 1 – (sum of gains of individual loops)

        + (sum of products of gains of all possible combinations of two non – touching loops)

        - (sum of products of gains of all possible combinations of three non – touching loops)

      + ...

    Δ_i = value of Δ when terms of Δ touching i^{th} forward path are removed

It may appear complicated but after a little bit of practice, this proves to be a very easy way of finding overall TF. The method of finding TF using MGF consists of:

1) Identify and write down all forward paths to get value of N.
2) Write down forward path gains F_1, F_2, ..., F_n.
3) Identify all loops and write them down.
4) Identify doublets, triplets, quadruplets of non – touching loops.
5) Write down determinant.
6) Write down each Δ_i by removing terms common with F_i.
7) Use MGF.

Example: Multiple Forward Paths

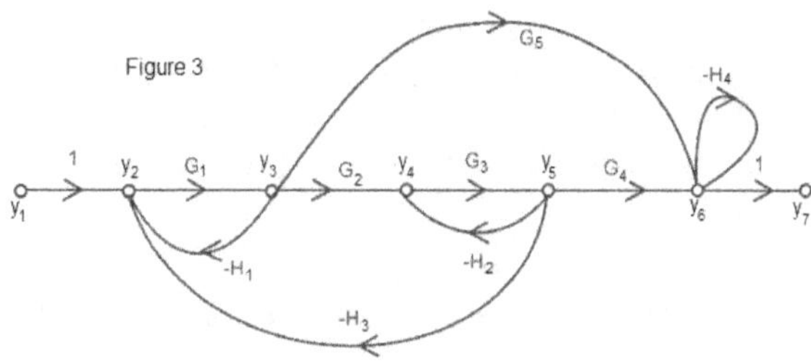

Find TF $\frac{y_7}{y_1}$ in the SFG shown below.

Solution

Forward Paths

1) $y_1 - y_2 - y_3 - y_4 - y_5 - y_6 - y_7;$ $\qquad F_1 = G_1 G_2 G_3 G_4$

2) $y_1 - y_2 - y_3 - y_6 - y_7;$ $\qquad F_2 = G_1 G_5$

   $\therefore N = 2$

Loops

1) $y_2 - y_3 - y_2$ $\qquad ; \qquad L_1 = -G_1 H_1$

2) $y_2 - y_3 - y_4 - y_5 - y_2$ $\qquad ; \qquad L_2 = -G_1 G_2 G_3 H_3$

3) $y_4 - y_5 - y_4$ $\qquad ; \qquad L_3 = -G_3 H_2$

4) $y_6 - y_6$ $\qquad ; \qquad L_4 = -H_4$

Pairs of two non touching loops: $(L_1, L_3), (L_1, L_4), (L_2, L_4), (L_3, L_4)$

Triads of three non touching loops: (L_1, L_3, L_4)

$$\therefore \Delta = 1 - (L_1 + L_2 + L_3 + L_4) + (L_1 L_3 + L_1 L_4 + L_2 L_4 + L_3 L_4) - L_1 L_3 L_4$$

Or $\Delta = -1 + G_1 H_1 + G_1 G_2 G_3 H_3 + G_3 H_2 + H_4 + G_1 G_3 H_1 H_2 + G_1 H_1 H_4 + G_1 G_2 G_3 H_3 H_4 + G_3 H_2 H_4 +$

$$G_1 G_3 H_1 H_2 H_4$$

$\Delta_1 =$ part of Δ non touching with $F_1 = 1$

$\Delta_2 =$ part of Δ non touching with $F_2 = 1 + G_3 H_2$

$$\therefore \frac{y_7}{y_1} = \frac{F_1 \Delta_1 + F_2 \Delta_2}{\Delta}$$

Home Work 14

1) Find the TF $\frac{y_2}{y_1}$ for the SFG shown in Figure 3 in previous example.

2) Find TF $\frac{y_5}{y_1}$ for the SFG shown below:

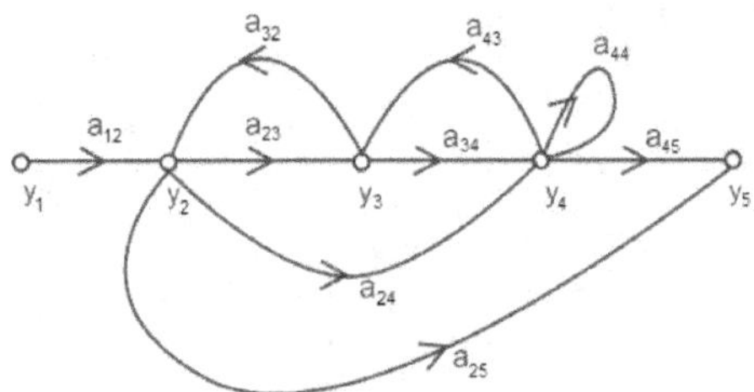

3) Drawn SFG for the block diagram shown below and thus find the TF $\frac{C(s)}{R(s)}$.

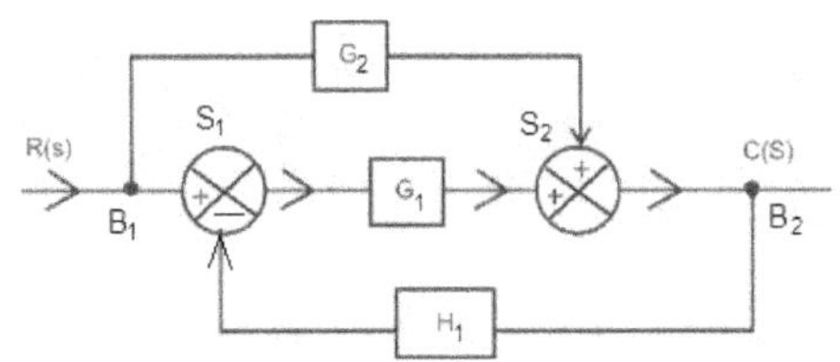

4) Draw SFG for the block diagram shown below and thus find the TF $\frac{C(s)}{R(s)}$.

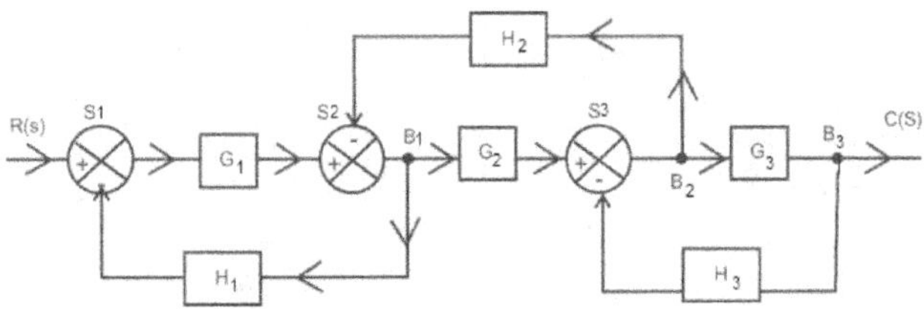

1.15. Time Response Analysis

Introduction

First step in analyzing / designing a control system is to derive a *mathematical model* of the system. Once such a model is obtained, several "what – if scenarios" may be studied using the model instead of actual experiments and this process is termed often as *simulation.*

Using the results of simulation, we may *compare* the *response* of one system with the other when both are subjected to *"demanding" input.* This process of comparison allows one to choose one particular system configuration over the other.

Test Signals

In practice, we do not know the input signal to the control system ahead of time since such input is typically random in nature. So, to *mimic* demanding inputs we use a few *test Signals* and compare the responses of *candidate system* when these test signals are input to the systems.

The commonly used test input signals are *step* function, *ramp* function, *acceleration* function, *impulse* function, *sinusoid* function and *white noise*.

Unit Impulse Function

Impulse function may be used to represent *sudden, instantaneous,* and *drastic* change. Examples where such situation occurs includes for example, hitting the system with a hammer to give a displacement input, breaking an automobile momentarily and sudden maneuver of a plane. mathematically unit impulse function is defined as:

$$h(t) = \begin{cases} \lim_{t_o \to 0} \dfrac{1}{t_o}, & 0 < t < t_o \\ 0 & , t < 0, t_o < t \end{cases}$$

Through the function appears a little formidable, its Laplace transform is very simple:

$$H(s) = L[h(t)] = 1.$$

Unit Step Function

Step function is used to represent *sudden and persistent change.* Examples in which such situations occur include placing a weight on weighing system and letting it "settle" down, pressing acceleration pedal of a car and keeping it pressed, and turning on outlet value in a liquid level system and keeping it open. Mathematically, unit step function is defined as:

$$u_s(t) = \begin{cases} 0, & t < 0 \\ 1, & \text{otherwise} \end{cases}$$

Laplace transform of u_s(t) is given as : $L[u_s(t)] = U_s(s) = \dfrac{1}{s}$

Unit Ramp Function

Ramp function may be used to represent *sudden, persistent and uniformly increasing* change. Mathematically, unit ramp function is defined as:

$$u_r(t) = \begin{cases} 0, & t < 0 \\ t, & \text{otherwise} \end{cases}$$

Laplace transform of u_r(t) is: U_r(s) = $L[u_r(t)] = \dfrac{1}{s^2}$

Unit Acceleration or Unit Parabolic Function

Acceleration input is used to represent *faster-than-ramp changes.* Mathematically unit acceleration function is defined as:

$$u_a(t) = \begin{cases} 0, & t < 0 \\ \dfrac{t^2}{2}, & \text{otherwise} \end{cases}$$

Laplace transform of $u_a(t)$ is defined as $U_a(s) = L[u_a(t)] = \dfrac{1}{s^3}$

Unit Sinusoidal Function

Sinusoidal input is used to represent *constantly fluctuating* input. This input is defined as:

$$f(t) = \begin{cases} 0, & t < 0 \\ \sin \omega t, & \text{otherwise} \end{cases}$$

Here ω is the *frequency* of signal. Its Laplace transform is:

$$F(s) = L[f(t)] = \frac{A\omega}{s^2 + \omega^2}.$$

White Noise Function

White noise is a random signal having equal intensity at different frequencies. such as input may be used to study response *of system to serially uncorrelated random input*. A white noise signal, denoted by $x(t)$ where samples of $x(t)$ are statistical uncorrelated, may be considered "noise" if covariance function is

$$Cov(x_i, x_j) = \begin{cases} \sigma^2, & i = j \\ 0, & i \neq j \end{cases}$$

We do NOT study white Noise signal: it is studied in a separate course on stochastic systems.

If the inputs considered in (i) to (v) in previous two pages have a magnitude other than unity, the word "Unit" may be omitted. For example, an impulse input of magnitude *A* may be written as "*Ah(t)*." In this context "*A*" may be taken as the magnitude of the test signal. We will study step-response, impulse-response and ramp-response in what follows. *Sinusoidal response* is studied in a separate section on *Frequency Response*.

We study the response of a typical first order system and a typical second order system when they are subjected to unit impulse input, unit step input, and unit ramp input.

1.16. First Order System

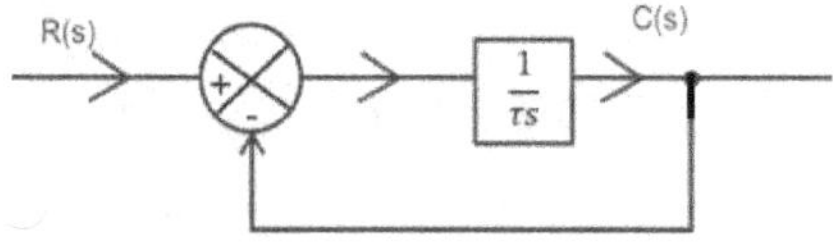

A typical first order system is shown in the figure beside. The input-output relationship of a first order system is given by,

$$\frac{C(s)}{R(s)} = \frac{1}{1 + \tau s} \;.\Rightarrow C(s) = \frac{1}{1 + \tau s} R(s)$$

Unit Impulse Input

$$r(t) = h(t) = \text{unit impulse input}$$

Laplace transform of $r(t) = R(s) = L[h(t)] = 1.$

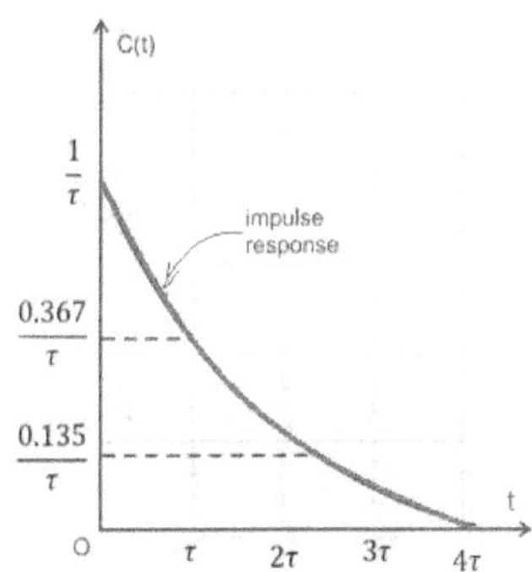

Thus, when $r(t) = h(t),\ \ R(s) = 1.$

$$\therefore C(s) = \frac{1}{1 + \tau s} = \frac{1}{\tau}\frac{1}{\left[s + \frac{1}{\tau}\right]}.$$

Taking inverse transform, $c(t) = \frac{1}{\tau}e^{-\left(\frac{t}{\tau}\right)}.$

The output of a first order system "decays" exponentially when the input is an impulse. The time constant of the system decides for how long the response c(t) remains non-zero. From the plot we see that after 4τ time period, practically the response c(t) is zero.

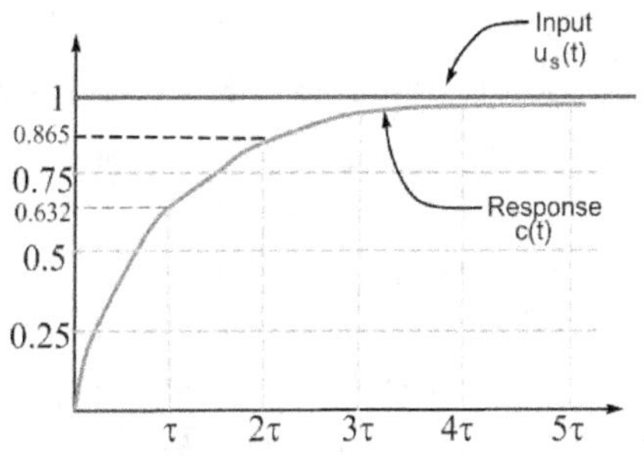

Unit Step Input

When the input to the system is unit step input,

$$r(t) = u_s(t) \Rightarrow R(s) = \frac{1}{s}$$

$$\therefore C(s) = \frac{1}{s(1 + \tau s)}$$

$$or \ C(s) = \frac{1}{s} - \frac{\tau}{1 + \tau s} = \frac{1}{s} - \frac{1}{s + \frac{1}{\tau}}$$

$$\therefore C(s) = 1 - e^{-\frac{t}{\tau}}$$

Step response of first order system	
t	C(t)
τ	0.632
2τ	0.865
3τ	0.95
4τ	0.962
5τ	0.993

Thus, we see that the output of a first order system to step input quickly raises to 63.2% of the stepe magnitude within the duration of 0 to τ seconds and then gradually approaches the input step.

It is clearly seen that $c(t) \to r(t)$ as $t \to \infty$ when $r(t) = u_s(t)$, since $\lim_{t \to \infty} c(t) = 1$.

1. Steady-state error is zero when a first order system is subjected to step input. First order system "tracks" step input.
2. When $t > 3\tau$, $r(t) - c(t) \le 1 - 0.95 \ or \ r(t) - c(t) \le 0.05$. Also,

    When $t > 4\tau$, $r(t) - c(t) \le 1 - 0.982 \ or \ r(t) - c(t) \le 0.02$.

    Thus, 2% settling time = 4τ

    5% settling time = 3τ

3. $\frac{dc(t)}{dt} = \frac{d}{dt}\left[1 - e^{-\frac{t}{\tau}}\right] = \frac{1}{\tau} e^{-t/\tau}$. Thus, derivative of step response is impulse response.

Unit Ramp Input

$$r(t) = u_r(t) \Rightarrow R(s) = \frac{1}{s^2}$$

$$\therefore C(s) = \frac{1}{s^2(1 + \tau s)} = \frac{1}{s^2} - \frac{\tau}{s} + \frac{T\tau^2}{\tau s + 1}$$

$$\Rightarrow C(t) = t - \tau + \tau e^{-(t/\tau)}$$

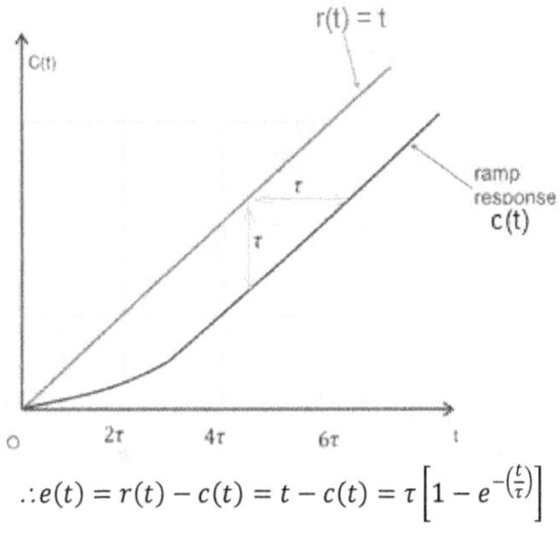

$$\therefore e(t) = r(t) - c(t) = t - c(t) = \tau\left[1 - e^{-\left(\frac{t}{\tau}\right)}\right]$$

$\therefore \lim_{t\to\infty} e(t) = \tau$. There is steady – state error.

The "gap" between the input r(t) and output c(t), as $t \to \infty$, will be τ long time axis and also along c(t)-axis.

1. First order system cannot "track" ramp input

2. $\frac{dc}{d\tau} = \frac{d}{d\tau}\left[t - T + Te^{-t/\tau}\right] = 1 - e^{-t/\tau} = $ step response.

See the picture below to observe that the ramp, step, and impulse inputs as well as responses bear a derivative relationship.

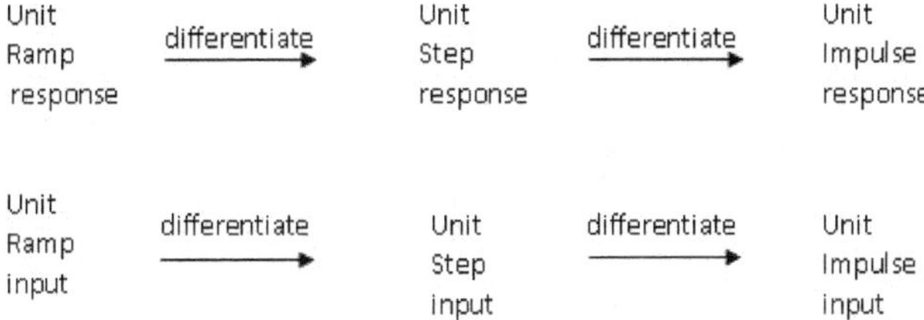

The response to the derivative of an input signal can be obtained by differentiating the response of the response of the system to the original signal. This is a property of linear-time-invariant systems. Linear time-varying systems and non-linear Systems do NOT posses this property.

Example: First Order System

A thermometer acting as a first order system is initially at a temperature of 35° and is then suddenly subjected to a temperature of 110 °. After 8 seconds, the thermometer indicates a temperature of 75 °. Calculate the time constant of the thermometer.

<u>Solution</u>: Let r(t) be the temperature of the medium and let c(t) be the temperature indicated by thermometer.

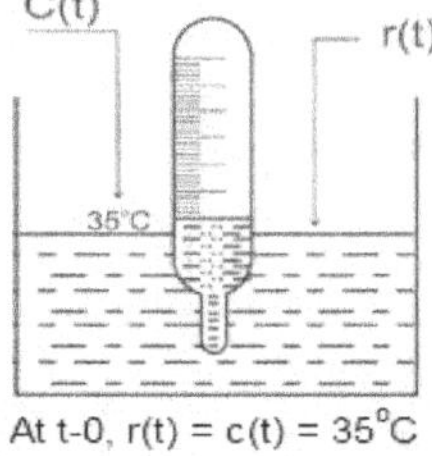

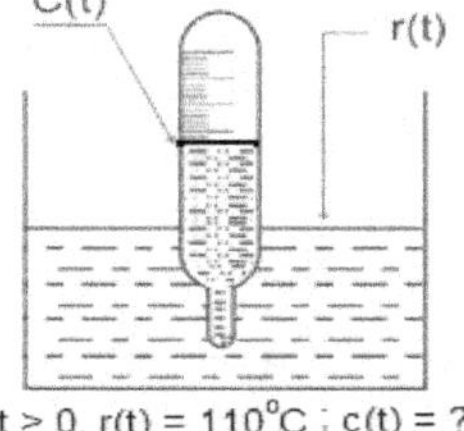

The initial condition and temperature input are shown pictorially above. Initially the temperature shown in the thermometer and the temperature of the medium are at 35 °C as shown in the left hand side of the picture above. The thermometer is then suddenly immersed into a medium with temperature of 110 °C.

Since temperature suddenly changed and remained at $110°$ it is a step input.

$$\frac{C(s)}{R(s)} = \frac{1}{1 + \tau s} \Rightarrow \left[1 + \tau \frac{d}{dt}\right] c(t) = r(t) = 110$$

[why do we have to write this? since initial condition$\neq 0$]

Taking Laplace transforms on both sides,

$$C(s) + \tau[sC(s) - 35] = \frac{110}{s} \Rightarrow C(s) = \frac{35\tau s + 110}{s(1 + \tau s)}$$

$$\text{Or } C(s) = \frac{110}{s} - \frac{75\tau}{1 + \tau s} \Rightarrow c(t) = 110 - 75e^{-\left(\frac{t}{\tau}\right)}.$$

$$\text{Given, } t = 8s, \ c(t) = 75°C \Rightarrow 75 = 110 - 75e^{-\left(\frac{t}{\tau}\right)}$$

$$\Rightarrow \tau \approx 10.5 \text{ sec}.$$

Home Work 15

1) A balloon carrying a first order thermocouple with a 15 second time constant rises through atmosphere at 6 m/s. Assume that the temperature varies with altitude at 0.15°C/30m. The balloon radios temperature and altitude reading back to the ground. At 3000m, the balloon says the temperature is 0°C. What is the true altitude at which 0°C occurs?

2) The temperature of a furnace is increasing at a rate of 0.1 °C/s. What is the maximum permissible time constant of a first order instrument that can be used to measure the temperature so that the temperature is read with a maximum error of 5 °C?

3) The following temperature-time data were recorded from a process:

Time (sec)	0	5	10	15	20	25	30	35	40	45	50
Temp ($^\circ$C)	20	85	130	160	185	190	195	197	200	202	205

From the plot, determine the time constant for the system. Also, write an equation assuming first order process.

4) The transfer function of a system is $G(s) = \dfrac{100}{(s+1)(s+100)}$. Find the approximate settling time for a 2% citerian, when subjected to unit step input. [Hint: read about dominant pole]

1.17. Second Order Systems

We consider the second order system shown in figure as an example and discuss second order systems.

The closed loop TF of this system is given by:

$$\frac{C(s)}{R(s)} = \frac{\overset{Zeroes\ :Nil}{K}}{\underset{Poles}{(Js^2 + Bs + K)}}.$$

Poles of the TF decide the transient response and stability of TF, so we write the <u>characteristic equation.</u>

$$Js^2 + Bs + K = 0 \text{ whose roots are: } s_1 = \frac{-B+\sqrt{B^2-4JK}}{2J} \text{ and } s_2 = \frac{-B-\sqrt{B^2-4JK}}{2J}$$

depending on the discriminant we say that

If $B^2 - 4JK < 0$ imaginary roots $\Rightarrow$ underdamped system

If $B^2 - 4JK = 0$ real, repeated roots $\Rightarrow$ critically damped

If $B^2 - 4JK > 0$ real district roots $\Rightarrow$ overdamped system

To generalize, we write the TF as:

$$\frac{C(s)}{R(s)} = \frac{K}{(Js^2+Bs+K)} = \frac{\left(\frac{K}{J}\right)}{s^2+\frac{B}{J}s+\frac{K}{J}} = \frac{\omega_n^2}{s^2+2\varepsilon\omega_n s+\omega_n^2}$$

Where $\omega_n^2 = \dfrac{k}{J}$ and $\varepsilon = \dfrac{B}{2\sqrt{JK}}$.

- ε and ω_n describe this system completely

- In critically damped system, $B^2 = 4JK \Rightarrow B = 2\sqrt{JK} = B_c$

- $\varepsilon = \dfrac{\text{Actual damping in system}}{\text{critical damping}} = \dfrac{B}{B_c}$

- If B is very small the response of the system is very oscillatory.

In terms of ε and ω_n, we say:

1. $\varepsilon < 1 \Rightarrow \begin{cases} S_1 = -\varepsilon\omega_n + j\omega_n\sqrt{1-\varepsilon^2} \\ S_2 = -\varepsilon\omega_n - j\omega_n\sqrt{1-\varepsilon^2} \end{cases} \Rightarrow$ under-damped system oscillatory response

 Critically damped system

2. $\varepsilon = 1 \Rightarrow \quad \underbrace{S_1 = S_2 = -\omega_n}_{\substack{This\ gives\ fastest \\ non-oscillatory\ response}} \Rightarrow$ non-oscillatory, non-overshooting system

 overdamped system

3. $\varepsilon > 1 \Rightarrow \begin{cases} S_1 = -\varepsilon\omega_n + \omega_n\sqrt{\varepsilon^2-1} \\ S_2 = -\varepsilon\omega_n + \omega_n\sqrt{\varepsilon^2-1} \end{cases} \Rightarrow$ non-oscillatory, $\varepsilon \Rightarrow$ sluggish response

Unit Step Response

We study unit step response of the second order system under the three cases (i), (ii), and (iii) listed above.

Under Damped: $\varepsilon < 1$

When the damping ratio is less than 1, the system response is oscillatory. We know that,

$$\varepsilon < 1 \Rightarrow \begin{cases} S_1 = -\varepsilon\omega_n + j\omega_n\sqrt{1-\varepsilon^2} \\ S_2 = -\varepsilon\omega_n - j\omega_n\sqrt{1-\varepsilon^2} \end{cases} \text{ and}$$

$$C(s) = \frac{\omega_n^2}{s(s^2 + 2\varepsilon\omega_n s + \omega_n^2)} = \frac{A}{s} + \frac{Bs + C}{s^2 + 2\varepsilon\omega_n s + \omega_n^2}$$

$$\Rightarrow A(s^2 + 2\varepsilon\omega_n s + \omega_n^2) + s(Bs + c) \equiv \omega_n^2 \ \forall\, s.$$

$$\textit{Equate Coeficcients of } \begin{cases} s^2 \text{ to get} & A + B = 0 \\ s^1 \text{ to get } 2\varepsilon\omega_n A + C = 0 \\ s^0 \text{ to get} & A = 1 \end{cases} \Rightarrow \begin{array}{l} A = 1 \\ B = -1 \\ C = -2\varepsilon\omega_n \end{array}$$

$$\therefore C(s) = \frac{1}{s} - \frac{s + 2\varepsilon\omega_n}{s^2 + 2\varepsilon\omega_n + \omega_n^2} = \frac{1}{s} - \frac{s + \varepsilon\omega_n}{s^2 + 2\varepsilon\omega_n + \omega_n^2} - \frac{\varepsilon\omega_n}{(s^2 + 2\varepsilon\omega_n + \omega_n^2)}$$

NOTE: $\mathcal{L}^{-1}\left[\frac{s+\varepsilon\omega_n}{s^2+2\varepsilon\omega_n+\omega_n^2}\right] = \mathcal{L}^{-1}\left[\frac{s+\varepsilon\omega_n}{(s+\varepsilon\omega_n)^2+\omega_d^2}\right] = e^{-\varepsilon\omega_n t}\cos(\omega_d t)$ where $\omega_d = \omega_n\sqrt{1-\varepsilon^2}$

And $\mathcal{L}^{-1}\left[\frac{\varepsilon\omega_n}{s^2+2\varepsilon\omega_n+\omega_n^2}\right] = \mathcal{L}^{-1}\left[\frac{\varepsilon\omega_n\cdot\omega_d}{\omega_d[s^2+2\varepsilon\omega_n+\omega_n^2]}\right] = \frac{\varepsilon}{\sqrt{1-\varepsilon^2}}e^{-\varepsilon\omega_n t}\sin(\omega_d t)$

$$\therefore c(t) = 1 - e^{-\varepsilon\omega_n t}\left[\cos(\omega_d t) + \frac{\varepsilon}{\sqrt{1-\varepsilon^2}}\sin(\omega_d t)\right]$$

$$\text{Or } c(t) = 1 - \frac{e^{-\varepsilon\omega_n t}}{\sqrt{1-\varepsilon^2}}\left[\overbrace{\sqrt{1-\varepsilon^2}}^{\sin\phi}\cdot\cos(\omega_d t) + \overbrace{\varepsilon}^{\cos\phi}\cdot\sin(\omega_d t)\right] = 1 - \frac{e^{-\varepsilon\omega_n t}}{\sqrt{1-\varepsilon^2}}\sin(\omega_d t + \phi)$$

Note

1. $\lim_{t \to \infty} c(t) = 1 \Rightarrow \lim_{t \to \infty} error = 0$

2. If $\varepsilon \to 0$, $c(t) = 1 - \cos(\omega_d t)$

3. When $\varepsilon \in [0.5, 0.8]$ very fast response than critically damped or over-damped response.

4. Two system having same ε but different ω_n show same overshoot. Such systems have same relative stability.

Critically Damped: $\varepsilon = 1$

Critically damped system have unity damping factor.

$$\varepsilon = 1 \Rightarrow C(s) = \frac{\omega_n^2}{s(s + \omega_n)^2} = \frac{A}{s} + \frac{B}{(s + \omega_n)} + \frac{C}{(s + \omega_n)^2}$$

$$\Rightarrow A(s + \omega_n)^2 + Bs(s + \omega_n) + cs \equiv \omega_n^2 \; \forall s$$

$$\begin{array}{ll}
s = 0 & \Rightarrow A = 1 \\
\Rightarrow s = -\omega_n & \Rightarrow C = -\omega_n \\
\text{Equate coeff.of } s^2 & \Rightarrow B = -1
\end{array}$$

$$\therefore C(s) = \frac{1}{s} - \frac{1}{s + \omega_n} - \frac{\omega_n}{(s + \omega_n)^2} \Rightarrow c(t) = 1 - \underbrace{e^{-\omega_n t}}_{term \; 1} - \underbrace{\omega_n t e^{-\omega_n t}}_{term \; 2}$$

$$\Rightarrow c(t) = 1 - e^{-\omega_n t} \, (1 + \omega_n t)$$

"Term 1" in response decays exponentially so does term 2, the little slowly since it involves a "t" in it.

$$\lim_{t \to \infty} c(t) = 1$$

Over Damped: $\varepsilon > 1$

An over-damped system has damping ratio greater than unity.

$$\varepsilon > 1 \Rightarrow \begin{cases} s_1 = -\varepsilon\omega_n + \omega_n\sqrt{\varepsilon^2 - 1} \\ s_2 = -\varepsilon\omega_n + \omega_n\sqrt{\varepsilon^2 - 1} \end{cases}$$

$$c(s) = \frac{\omega_n t}{s(s - s_1)(s - s_2)} \quad \text{when } R(s) = \frac{1}{s}$$

After a bit of algebra, we split $C(s)$ as:

$$C(s) = \frac{1}{s} + \frac{1}{2\sqrt{\varepsilon^2 - 1}[-\varepsilon + \sqrt{\varepsilon^2 - 1}]}\frac{1}{(s - s_1)} + \frac{1}{2\sqrt{\varepsilon^2 - 1}[\varepsilon + \sqrt{\varepsilon^2 - 1}]}\frac{1}{(s - s_2)}$$

$$\text{So, } c(t) = 1 + \underbrace{\frac{e^{s_1 t}}{2\sqrt{\varepsilon^2-1}[-\varepsilon+\sqrt{\varepsilon^2-1}]}}_{term \; 3} + \underbrace{\frac{e^{s_2 t}}{2\sqrt{\varepsilon^2-1}[\varepsilon+\sqrt{\varepsilon^2-1}]}}_{term \; 4}$$

Note

1. s_2 is always less than zero. So $e^{s_2 t}$ decays as $t \to \infty$.

2. When damping in the system is very large, $\varepsilon \gg 1$ then $|s_2| \gg |s_1|$. Then term 4 decays much faster than term 3. So, we may omit the term 3 in an approximation. . We say that s_1 is the *dominant pole*. In such situations we may write:

$$c(t) \approx 1 - e^{-(+\varepsilon-\sqrt{\varepsilon^2-1})\omega_n t}$$

MATLAB is a very useful tool for analyzing / stimulating dynamic systems and a control engineer must familiarize himself herself with analysis / stimulation tool. A MATLAB program is shown on page 78 and unit step response of a second order system for various values of damping ratio is shown in Figure A on page 80 .

- The family of curves are drawn for $\omega_n = 2$ rad/sec and various values of ε
- For other value of ω_n also, the family of curves will look "similar".
- Two second-order system having the same ε but different ω_n will exhibit the same overshoot and same obcilliatory pattern. Such system are said to have same *Relative Stability.*
- From response curve on page 80, we see that ε in $[0.5, 0.8]$ gives faster response [i.e. the system gets close to final value] than a critically damped or overdamped system.
- Among the systems responding to without oscillation, critically damped system exhibits fastest response.
- An over damped system is sluggish in responding to inputs
- These observations apply to the transfer functions shown in green dashed box on Page 75 and may be different for other transfer functions.

A Matlab Script

```
clc
clear
close all
label_marker=0;
damping_ratio=0;
%damping_ratio=1;
if damping_ratio==0;
    zeta_range=[0:0.3:1,1.2:0.4:2.4];
else
    zeta_range=0.5;
```

```matlab
end
for zeta=zeta_range;
    wn=2;
    t=[0:0.1:6];
    u=ones(size(t));
    sys=tf(wn^2,[1 2*wn*zeta wn^2]);
    y1=lsim(sys,u,t);
    plot(wn*t,y1);
    hold on;
    label_marker=0;
    if label_marker==0
        gtext(strcat('\zeta=',num2str(zeta,2)));
        label_marker=1;
    elseif label_marker==1
        label_marker=0;
    end
end
grid on
xlabel('\omega_nt---->')
ylabel('c(t)---->');
if damping_ratio==0
    title('Second Order System: Effect of damping Ratio \zeta');
else
    wnt=wn*t;

    title('Response Specifications of Second order system');
    line1=1.02*ones(size(y1));
    line2=0.98*ones(size(y1));
    plot(wnt,line1,'r--',wn*t,line2,'--');
    gtext('1.02 line');
    gtext('0.98 line');

    %2% settling time
    y2=0:0.14:0.98;
    x2=8*ones(size(y2));
    plot(x2,y2,'b-.')
```

```matlab
gtext('2% Settling Time')

%rise time
y2=0:0.1:1;
x2=2.4*ones(size(y2));
plot(x2,y2,'b-.')
gtext('Rise Time')

%peak time
y2=0:0.01:1.163;
x2=3.6*ones(size(y2));
plot(x2,y2,'b-.')
gtext('Peak Time')

end
```

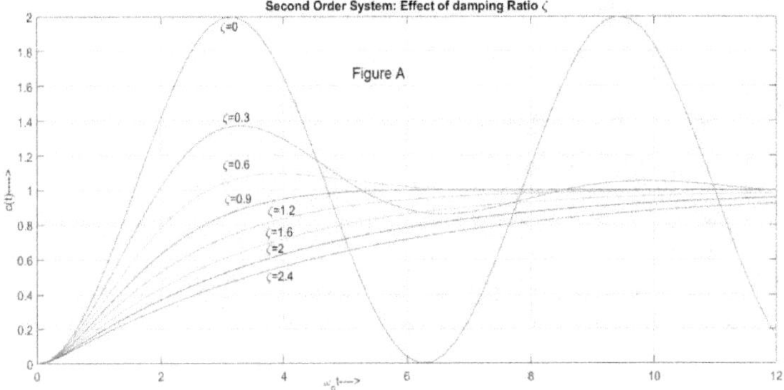

Figure 1: Response for Various Damping Ratios

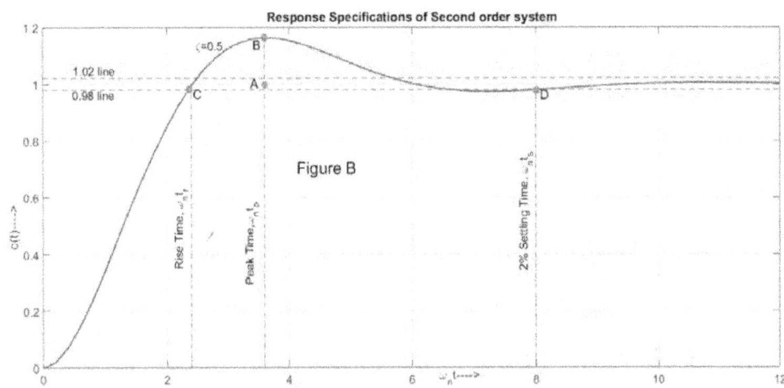

1.18. Transient Response Specifications

The response of the LTI system to any arbitrary input can be mathematically deducted from the response of the system to unit step input. Thus, performance characteristics of a LTI control systems are specified in terms of transient response to a unit step input. For easy comparison, the system under consideration is assumed to be at rest initially output and all time derivatives are assumed to be zeros.

Plot in Figure B shows the transient response of a typical control system which "oscillates" before attaining steady state. Such Response is common feature of a practical control system. The following specifications are commonly used in describing transient response.

Risk Time, t_r

1. It is the time required for the response to rise from 0% to 100% of the steady-state value. For over-damped systems 10% to 90% rise time is commonly used.

   For under-damped system, we have,

   $$c(t) = 1 - \frac{e^{-\varepsilon\omega_n t}}{\sqrt{1-\varepsilon^2}} \sin(\omega_d + \phi) \text{ where } \tan\phi = \frac{\sqrt{1-\varepsilon^2}}{\varepsilon}$$

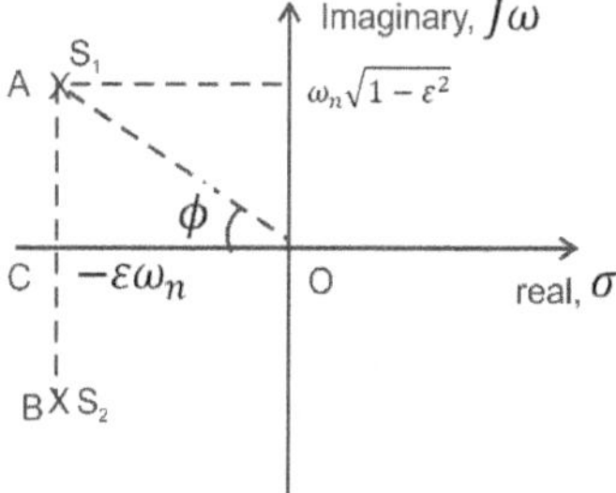

At $t = t_r$ we have $c(t_r) = 1$

$$\Rightarrow 1 = 1 - \frac{e^{-\varepsilon\omega_n t_r}}{\sqrt{1-\varepsilon^2}} \sin(\omega_d t_r + \phi)$$

$$\Rightarrow \sin(\omega_d t_r + \phi) = 0 \Rightarrow \omega_d t_r + \phi = \pi$$

$$\Rightarrow t_r = \frac{\pi - \phi}{\omega_d}$$

A geometric interpretation is shown in Figure where s_1, s_2 are shown as poles ×.

Peak Time t_p

The time at which the response curve c(t) attains maximum value is termed *peak time.*

$$\text{At } t = t_p, \quad \frac{dc}{dt}\bigg|_{t=t_p} = 0$$

$$\Rightarrow \frac{e^{-\varepsilon\omega_n t_p}}{\sqrt{1-\varepsilon^2}}\cos(\omega_d t_p + \phi) * \omega_d = \frac{\sin(\omega_d t_p + \phi)}{\sqrt{1-\varepsilon^2}}\varepsilon\omega_n e^{-\varepsilon\omega_n t_p}$$

$$\Rightarrow \tan(\omega_d t_p + \phi) = \frac{\sqrt{1-\varepsilon^2}}{\varepsilon} = \tan\phi \Rightarrow \frac{\sin(\omega_d t_p + \phi)}{\cos(\omega_d t_p + \phi)} = \frac{\sin\phi}{\cos\phi}$$

$$\Rightarrow \sin(\omega_d t_p + \phi)\cos\phi - \cos(\omega_d t_p + \phi)\sin\phi = 0$$

$$\Rightarrow \sin(\omega_d t_p) = 0 \Rightarrow \omega_d t_p = \pi \Rightarrow t_p = \frac{\pi}{\omega_d}$$

Peak Over Shoot M_p

Overshoot at any time is $|c(t) - 1|$. Maximum value of overshoot, M_p is defined as the peak overshoot. At $t = t_p$, $c(t) = AB$, as seen in Figure B on page 80, or

$$c(t_p) = AB = 1 - \frac{e^{-\varepsilon\omega_n t_p}}{\sqrt{1-\varepsilon^2}}\sin(\omega_d t_p + \phi)$$

$$\Rightarrow AB = 1 + \frac{e^{-\varepsilon\omega_n t_p}}{\sqrt{1-\varepsilon^2}}\sin\phi \quad \because t_p = \frac{\pi}{\omega_d} \text{ and } \sin(\pi + \phi) = -\sin\phi$$

$$\Rightarrow AB = 1 + e^{-\varepsilon\omega_n t_p} \quad \because \sin\phi = \sqrt{1-\varepsilon^2}.$$

$$\Rightarrow AB = 1 + e^{\frac{-\pi\varepsilon}{\sqrt{1-\varepsilon^2}}} = c(t_p) \quad \because t_p = \frac{\pi}{\omega_d} = \frac{\pi}{\omega_n\sqrt{1-\varepsilon^2}}$$

$$\% \ M_p = \frac{c(t_p)-1}{1} * 100 = 100 \ e^{\frac{-\pi\varepsilon}{\sqrt{1-\varepsilon^2}}}$$

$$\% \ \text{Max. overshoot} = 100 \ e^{\frac{-\pi\varepsilon}{\sqrt{1-\varepsilon^2}}}$$

Settling Time, t_s

2% settling time is the time after which the response curve lies within the 2% tolerance band about the steady state value. Usually 2% settling time or 5% settling time is defined. For 2% settling time t_s, we use the definition that

$$1.02 < |c(t) - 1| < 0.98 \quad \forall t \geq t_s$$

Or for 5% settling time t_s,

$$1.05 < |c(t) - 1| < 0.95 \quad \forall t \geq t_s$$

It is left to the students to prove that

$$t_s \approx \frac{4}{\varepsilon\omega_n} \text{ for 2\% tolerance band, and}$$

$$t_s \approx \frac{3}{\varepsilon\omega_n} \text{ for 5\% tolerance band}$$

When a system does NOT exhibit required response, it may be possible to use a *"compensator"* and achieve desired response.

We consider one particular system as an example and consider two particular control schemes. *Proportional-Derivative, Proportional- Integral.*

Proportional Derivative (PD) Controller

Consider a system which has the Transfer function

$$G_p(s) = \frac{\omega_n{}^2}{s(s + 2\varepsilon\omega_n)}$$

Along with a controller that has transfer function as shown in Fig. below

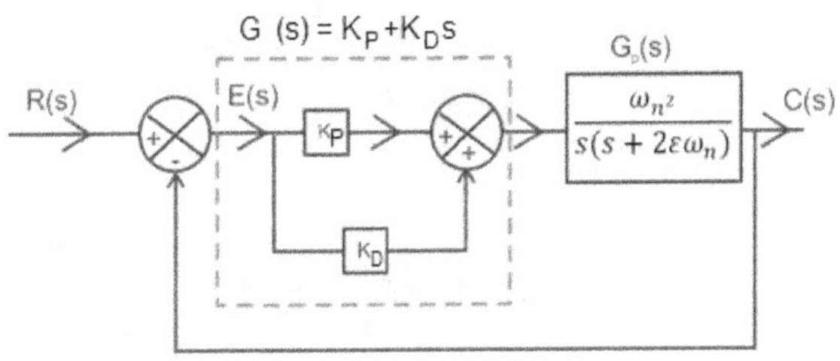

The overall TF of this block diagram is

$$\frac{c(s)}{R(s)} = \frac{\omega_n^2(K_P + K_{Ds})}{s^2 + (2\varepsilon\omega_n + \omega_n^2 K_D)s + \omega_n^2 K_P}$$

If we want to achieve poles corresponding to an arbitrary natural frequency λ and a damping ration Ψ, we note that

$$\lambda = \sqrt{constant\ term} = \omega_n\sqrt{K_P} \quad \text{and}$$

$$\Psi = \frac{coefficient\ of\ s}{2\sqrt{constant - term}} = \frac{2\varepsilon + \omega_n K_D}{2\omega_n\sqrt{K_P}}$$

By properly choosing K_P and K_D, it may be possible to achieve desired values of λ and Ψ.

PD controller in general increases damping

- The PD controller adds a zero at s = - K_P/K_D .
- The PD controller in general increases damping.
- Increasing the K_D value in general leads to decrease the peak overshoot, the rise time, and the settling time.
- PD control is **anticipatory control**
- PD control will have an effect only if the steady state error is varying. If steady state error is constant, PD control cannot kill it.
- PD controller introduces a **corner frequency** in frequency response curve
- PD controller acts as *a high pass filter*
- PD controller with generally increase **bandwidth** and reduce the **rise time**. Improves **gain margin, phase margin**.

It may be noted here that judicious choice of K_P and K_D depends on the particular system being studied and must involve through simulation. MATLAB comes in very handy at this stage.

It may be mentioned in passing that implementing a PD controller may require a relatively large capacitor in circuit implementation.

Proportional Integral Control (PI)

We have seen that PD controller cannot "kill" a constant steady state error. The PI controller, shown in figure below can achieve this objective when properly designed.

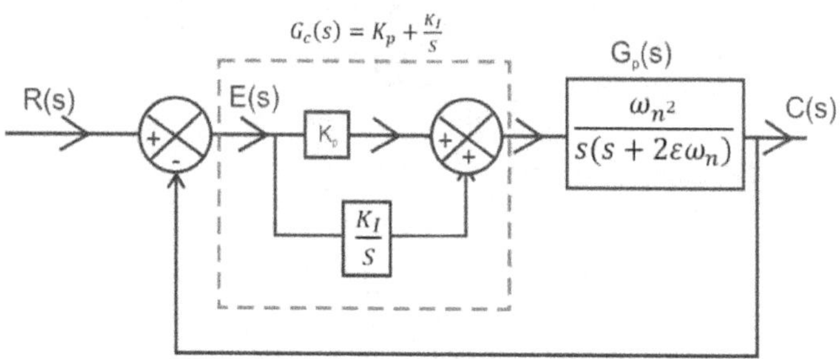

The over all TF of the system is: $\dfrac{c(s)}{R(s)} = \dfrac{\omega_n{}^2(k_p s + k_I)}{s^3 + 2\varepsilon\omega_n s^2 + \omega_n{}^2 k_p s + \omega_n{}^2 k_I}$

In this case, the analysis is a bit more complicated since the characteristic equation now became cubic.

- Implementing PI controller will involve adding a zero at $s = -\dfrac{K_P}{K_I}$.
- The system type is increased by 1
- Constant steady state error can be "killed" using PI controller provided compensated system remains stable
- If K_P and K_I are not properly chosen, the compensated system may go unstable
- PI control improves damping and reduces peak overshoot
- *PI control increases rise – time*
- PI control decreases band – width
- May be useful in filtering high – frequency noise
- Improves gain margin and phase margin
- In general, the implementation of the PI controller yields a larger capacitor value. Design process to achieve smaller capacitors is more difficult than in the case of a PD controller.

1.19. Steady State Errors in Unity Feedback Control Systems

There may be steady state errors due to changes in the command signal or the steady state errors may be due to imperfections in the system components such as static friction, backlash, hysteresis, amplifier drift, and aging or deterioration of system components.

Present discussion concerns itself only steady state error arising out of incapability of the system to follow particular types of input and does NOT attempt to address imperfections mentioned in previous paragraph.

Most control systems show steady – state error in response to certain types of inputs. A system may have zero steady state error when subjected to a step input but the same system may exhibit non zero steady state error when reference is a ramp input. In situations of non – zero steady state error, modifying the open – loop structure of the system may be only way to achieve zero steady state error. Whether a given open loop system exhibits steady – state error for a given type of input depends on the *type of open – loop transfer function* of the system.

Type of System

Consider a system with unity feedback and the following open loop transfer function:

$$G(s) = \frac{K(1 + \tau_1 s)(1 + \tau_2 s) \dots \dots \dots (1 + \tau_m s)}{s^N (1 + T_1 s)(1 + T_2 s) \dots \dots \dots (1 + T_n s)}.$$

This system has N isolated integrations present in OLTF and is called *type-N system*. If N=1, we call it *type-1 system*, if N=2, we call it *type-2 system* etc.

Note that the "order" of the system is (N+n) where its "type" is N.

As the type number is increased, accuracy is improved; however, increasing the type number very much may make the system so sensitive that it may cause *stability* problem.

Design of controller involves compromise between steady state accuracy and relative stability, many times.

Notice that each of the terms $(1 + \tau_1 s), (1 + \tau_2 s), \dots \dots \dots (1 + \tau_m s)$ and $(1 + T_1 s)(1 + T_2 s) \dots \dots \dots (1 + T_n s)$ approaches unity as $s \to 0$. In such case, the open loop gain K is directly related to the steady state error.

Consider the system shown in block diagram below:

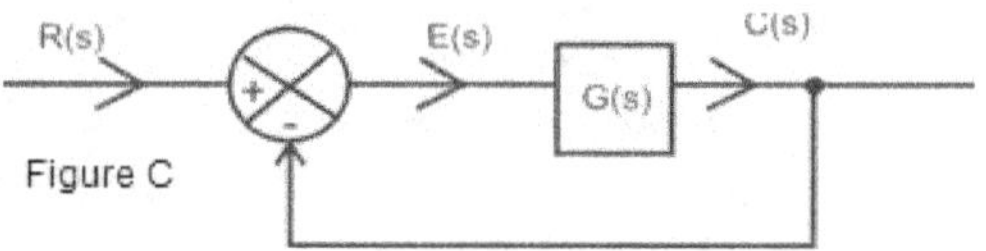

Figure C

The closed loop TF

$$\frac{C(s)}{R(s)} = \frac{G(s)}{1+G(s)}$$

and the TF between $e(t)$ and $c(t)$ is

$$\frac{E(s)}{R(s)} = \frac{1}{1+G(s)} \Rightarrow E(s) = \frac{1}{1+G(s)} R(s).$$

Therefore steady – state error is obtained as

$$e_{ss} = \lim_{t\to\infty} e(t) = \lim_{s\to o} sE(s) = \lim_{s\to o} \frac{sR(s)}{1+G(s)}.$$

Based on this discussion we define <u>static error constants</u> such that the higher the error constant, the smaller the steady state error.

In this discussion, we use the term "position" to mean the output of the system, we use the term "velocity" to mean the rate of the output (or first derivative), and we use the term "acceleration" to mean second derivative of output.

We define three constants

- Static Position error constant, K_P
- Static velocity error constant, K_v
- Static acceleration error constant, K_a

K_P gives and indication of steady state error (SSE) for unit step input; K_V gives an indication of SSE for ramp input; K_a gives an indication of SSE for acceleration input.

1. *Static Position Error Constant*

We define the static position error constant in the context of finding the SSE when the system is subjected to a unit step input.

If $R(s) = \frac{1}{s}$ in Figure C,

$$e_{ss} = \lim_{s\to o} \frac{sR(s)}{1+G(s)} = \lim_{s\to o} \frac{1}{1+G(s)} = \frac{1}{1+\lim_{s\to o} G(s)}$$

If we let $\qquad K_p = \lim_{s\to o} G(s), \qquad$ we see that $\qquad e_{ss} = \frac{1}{1+K_p}$

For Type-0 system:- $K_p = \lim_{s\to o} \dfrac{K(1+\tau_1 s)(1+\tau_2 s)\ldots\ldots(1+\tau_m s)}{s^N (1+T_1 s)(1+T_2 s)\ldots\ldots(1+T_n s)} = K$

$$\therefore e_{ss} = \frac{1}{1+K}$$

For Type-1 system:- $K_p = \lim_{s\to o} \dfrac{K(1+\tau_1 s)(1+\tau_2 s)\ldots\ldots(1+\tau_m s)}{s^N (1+T_1 s)(1+T_2 s)\ldots\ldots(1+T_n s)} = \infty$

$$\therefore e_{ss} = \frac{1}{1+K} = 0$$

We see that for Type-1 system, Type-2 System, and all higher order types of systems, $K_p = \infty$ and $e_{ss} = 0$. ***This indicates that one integrator is sufficient to remove steady state error for step input.***

2. Static Velocity Error Constant

This error constant is defined in the context of computing the SSE of the system subjected to ramp input.

$$\text{If } R(s) = \frac{1}{s^2} \text{ in Figure C,}$$

$$e_{ss} = \lim_{s \to o} \frac{sR(s)}{1 + G(s)} = \lim_{s \to o} \frac{s * \frac{1}{s^2}}{1 + G(s)} = \lim_{s \to o} \frac{1}{s + sG(s)} = \lim_{s \to o} \frac{1}{sG(s)}.$$

If we let $K_v = \lim_{s \to o} sG(s)$, we see that $e_{ss} = \dfrac{1}{K_v}$

The term "velocity error", e_{ss} here refers to steady-state error for a ramp input. Velocity error is not an error in velocity! It is an error in position due to ramp input: It is the ***tracking error*** while tracking a ramp input.

<u>For Type-0 system</u>:- $K_v = \lim_{s \to o} \dfrac{sK(1+\tau_1 s)(1+\tau_2 s)\ldots\ldots(1+\tau_m s)}{(1+T_1 s)(1+T_2 s)\ldots\ldots(1+T_n s)} = 0$

$$\therefore e_{ss} = \frac{1}{K_v} = \infty$$

<u>For Type-1 system</u>:- $K_\vartheta = \lim_{s \to o} \dfrac{sK(1+\tau_1 s)(1+\tau_2 s)\ldots\ldots(1+\tau_m s)}{s(1+T_1 s)(1+T_2 s)\ldots\ldots(1+T_n s)} = K$

$$\therefore e_{ss} = \frac{1}{K_v} = \frac{1}{K}$$

For Type-2 or 3 or higher type systems, $K_v = \infty \Rightarrow e_{ss} = 0$

3. Static Acceleration Error Constant

We define this constant in the context of find SSE when the system is subjected to a parabolic input.

For $r(t) = \frac{1}{2}t^2 \Rightarrow R(s) = \frac{1}{s^3}$ in Figure C.

$$e_{ss} == \lim_{s \to o} \frac{sR(s)}{1 + G(s)} = \lim_{s \to o} \frac{1}{s^2 G(s)} = \frac{1}{K_a}$$

where $\qquad K_a = \lim_{s \to o} s^2 G(s)$

<u>For Type-0 system</u>:- $K_a = \lim_{s \to 0} \dfrac{s^2 K(1+\tau_1 s)(1+\tau_2 s)\ldots\ldots(1+\tau_m s)}{(1+T_1 s)(1+T_2 s)\ldots\ldots(1+T_n s)} = 0 \Rightarrow e_{ss} = \infty$

For Type-1 system:- $K_a = \lim_{s \to 0} \dfrac{s^2 K(1+\tau_1 s)(1+\tau_2 s)\ldots\ldots(1+\tau_m s)}{s(1+T_1 s)(1+T_2 s)\ldots\ldots(1+T_n s)} = 0 \Rightarrow e_{ss} = \infty$

For Type-2 system:- $K_a = \lim_{s \to 0} \dfrac{s^2 K(1+\tau_1 s)(1+\tau_2 s)\ldots\ldots(1+\tau_m s)}{s^2(1+T_1 s)(1+T_2 s)\ldots\ldots(1+T_n s)} = K \Rightarrow e_{ss} = \frac{1}{K}$

For Type-3 system:- $K_a = \lim_{s \to 0} \dfrac{s^2 K(1+\tau_1 s)(1+\tau_2 s)\ldots\ldots(1+\tau_m s)}{s^3(1+T_1 s)(1+T_2 s)\ldots\ldots(1+T_n s)} = \infty \Rightarrow e_{ss} = 0$

For higher types of systems, $K_a = \infty \Rightarrow e_{ss} = 0$

Table 1: <u>Static Error Constants and Steady – State Errors</u>

System	Step input $r(t) = 1$	Ramp input $r(t) = t$	Acceleration input $r(t) = \dfrac{1}{2}t^2$
Type 0 system	$\dfrac{1}{1+K_p} = e_{ss}$	$e_{ss} = \infty$	$e_{ss} = \infty$
Type 1 system	0	$\dfrac{1}{K_v}$	∞
Type 2 system	0	0	$\dfrac{1}{K_a}$
Type 3	0	0	0

At the table above shows, for a Type 3 system, which has three integrators in series, the steady state error is zero. the zeros steady state error for step, ramp, and acceleration input may look very attractive. However, additional integrators cause additional stability problems! The design of a satisfactory system with more than two integrators in series in the feed forward path is generally not easy!

Example 1: Standard Second Order System

A closed loop system has $TF: \dfrac{c(s)}{R(S)} = \dfrac{36}{s^2+2s+36}$. Determine the rise time, peak time, peak overshoot, and 2% settling time of this system when $R(s) = \dfrac{1}{s}$.

<u>Solution</u>: Comparing this TF to standard form $\dfrac{\omega_n^2}{s^2+2\varepsilon\omega_n s+\omega_n^2}$, we see that $\omega_n = 6$ rad/s and $2\varepsilon\omega_n = 2 \Rightarrow \varepsilon = \dfrac{1}{\omega_n} = \dfrac{1}{6}$.

Therefore that $\omega_n = 6$ rad/s and $2\varepsilon\omega_n = 2 \Rightarrow \varepsilon\omega_n = 1 \Rightarrow \varepsilon = \dfrac{1}{\omega_n} = \dfrac{1}{6}$. Hence, we can see,

$$\omega_d = \omega_n\sqrt{1-\varepsilon^2} = 5.92 \text{ rad/sec and } \phi = tan^{-1}\left(\dfrac{\sqrt{1-\varepsilon^2}}{\varepsilon}\right) = 80.41° = 1.40 \text{ rad}$$

$\therefore$ Rise – time, $t_r = \dfrac{\pi-\beta}{\omega_d} = 0.29$ sec

Peak – time, $t_p = \dfrac{\pi}{\omega_d} = 0.53$ sec

% Peak overshoot $= 100e^{\frac{-\pi\varepsilon}{\sqrt{1-\varepsilon^2}}} = 58.8\%$

2% Settling time, $t_s = \dfrac{4}{\varepsilon\omega_n} = 4$ sec

Example 2: Error Constants

A closed- loop system has transfer function $\dfrac{c(s)}{R(s)} = \dfrac{\omega_{n}^{2}}{s^{2}+2\varepsilon\omega_{n}s+\omega_{n}^{2}}$ with $\omega_{n} = 6$ rad/sec and $\varepsilon = \dfrac{1}{6}$. Find the position, velocity, and acceleration error constants. Also find the steady state error in the output of this system when subjected to unit step input, Unit ramp input, and unit acceleration input.

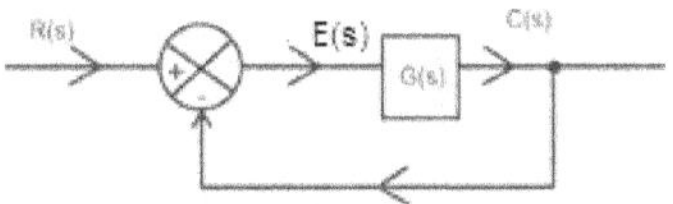

<u>Solution</u>:- We have $\dfrac{C(s)}{R(s)} = \dfrac{36}{s^{2}+2s+36} = \dfrac{G(s)}{1+G(s)}$ where $G(s)$ is the transfer function of the open loop system shown in figure. Here, we have assumed Unity, negative feedback

$$\frac{G(s)}{1+G(s)} \;=\; \frac{36}{s^{2}+2s+36} \Rightarrow (s^{2}+2s)G(s)+\overbrace{36G(s)}^{Cancel} = 36 + \overbrace{36G(s)}^{Cancel}$$

$$\Rightarrow G(s) = \frac{36}{s(s+2)}\,.$$

Thus, the type of the system is 1.

Position Error Constant, $K_{p} = \lim_{s\to 0} G(s) = \infty \Rightarrow e_{ss}$ for step input $= \dfrac{1}{1+K_{p}} = 0$

Velocity Error Constant, $K_{v} = \lim_{s\to 0} sG(s) = \dfrac{36}{2} = 18 \Rightarrow e_{ss}$ for ramp input $= \dfrac{1}{K_{v}} = \dfrac{1}{18}$

Acceleration Error Constant, $K_{a} = \lim_{s\to 0} s^{2}G(s) = 0 \Rightarrow e_{ss}$ for $r(t) = \dfrac{1}{K_{a}} = \infty$

<u>Consider the alternative way of finding position steady state error:</u>

Let $r(t) = 1 \Rightarrow r_{ss} =$ steady – state value of $r(t) = 1$ and $R(s) = \dfrac{1}{s}$

$\therefore c(s) = \dfrac{36}{s^{2}+2\varepsilon\omega_{n}s+\omega_{n}^{2}}$ and $c_{ss} =$ steady-state value of $c(t)$

$$= \lim_{s\to 0} sC(s) = \frac{36}{\omega_{n}^{2}} = 1$$

$\therefore e_{ss} =$ steady-state error $= r_{ss} - c_{ss} = 1 - 1 = 0$

in finding e_{ss} for step input both methods have same level of difficulty. But finding e_{ss} for ramp and acceleration input using alternate way is a bit more tedious! Think.

Example 3: Inner-Loop Control

Consider the system shown in Figure D. The damping ratio of the system is 0.158 and the undamped natural frequency is 3.16 rad/sec. To improve the relative stability, we employ techometre feedback as shown in Figure E. Determine the value of K_b so that the damping ratio of the closed loop system in Figure E is 0.5.

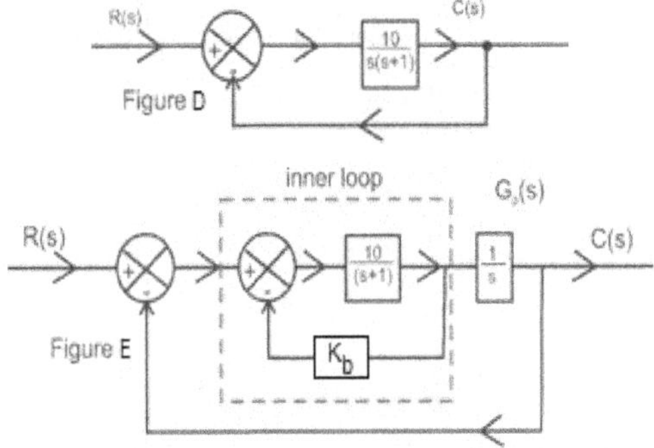

Solution:- This problem is really, really simple. Evaluate the overall transfer function for the system shown in figure E.

$$\frac{C(s)}{R(s)} = \frac{\overbrace{10}^{\text{Derive this and convince yourself}}}{s^2 + (1 + 10K_b)s + 10} = \frac{\omega_n{}^2}{s^2 + 2\varepsilon\omega_n s + \omega_n{}^2}$$

$$\Rightarrow \omega_n = \sqrt{10} = 3.16 \text{ rad/sec}$$

It is required that $\varepsilon = 0.5 \Rightarrow \frac{1+10K_b}{2\sqrt{10}} = 0.5 \Rightarrow K_b = \frac{\sqrt{10}-1}{10} = 0.216$

Note

(i) Knowing block diagram reduction and signal flow graphs is very essential. You may score lot of marks very easily. In the problem discussed just now, it is a matter of 1 minute to solve it after finding overall TF!

(ii) Noticed that by introducing an "inner loop", the damping ratio is increased from 0.158 to 0.5: a substantial improvement in damping ratio. System in Figure E is less oscillatory and settles down quickly. **System in figure A status in 8 seconds And system in figure B settled in** $\frac{8}{\sqrt{10}}$ **sec. Derive it for yourself!** To achieve this, we need to use a tachometer which involves an expense! In this type of situations, an *observer* may be designed to compute the speed.

Example 4: Finding Forward Gain and Feedback Loop Gain

Determine the value of K and α of the closed loop system in figure so that the maximum overshoot in unit step response is 25% and the peak time is 2 second. $J = 1$ Kg-m^2.

Solution

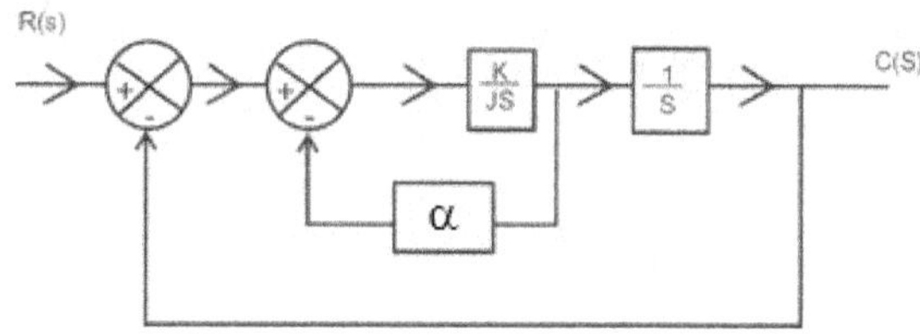

Closed loop transfer function is obtained as

$$\frac{c(s)}{R(s)} = \frac{K}{Js^2+K\alpha s+k} = \frac{K}{s^2+K\alpha s+K} \qquad \because J = 1\text{Kg-m}^2$$

$\therefore$ Comparing it with standard TF $\dfrac{\omega_n{}^2}{s^2+2\varepsilon\omega_n s+\omega_n^2}$, we see

$\omega_n = \sqrt{K}$ and $2\varepsilon\omega_n = k\alpha$.

Maximum overshoot, $M_p = 25\% = 100e^{\frac{-\pi\varepsilon}{\sqrt{1-\varepsilon^2}}} \Rightarrow \dfrac{\pi\varepsilon}{\sqrt{1-\varepsilon^2}} = 1.386 \Rightarrow \varepsilon = 0.404$

Peak-time $= t_p = 2 \text{ sec} = \dfrac{\pi}{\omega_d} \Rightarrow \omega_d = \dfrac{\pi}{2} = 1.57 \text{ rad/sec}$

$\therefore$ Undamped natural frequency, $\omega_n = \dfrac{\omega_d}{\sqrt{1-\varepsilon^2}} = \dfrac{1.57}{\sqrt{1-0.404^2}} = 1.72 \text{ rad/sec}$

Therefore, we obtain

$K = \omega_{n^2} = 1.72^2 = 2.95 \; N - M$

$\alpha = \dfrac{2\omega_{n^2}}{K} = \dfrac{2*0.404*1.72}{2.95} = 0.471 \text{ sec}$

Home Work 16

1) In experimentation with an unknown electrical circuit, the output voltage is obtained as shown in the plot becide when subjected to a 10 v step input. If it is known that the circuit is approximates a second order system, obtain an estimate of the transfer function of the circuit.

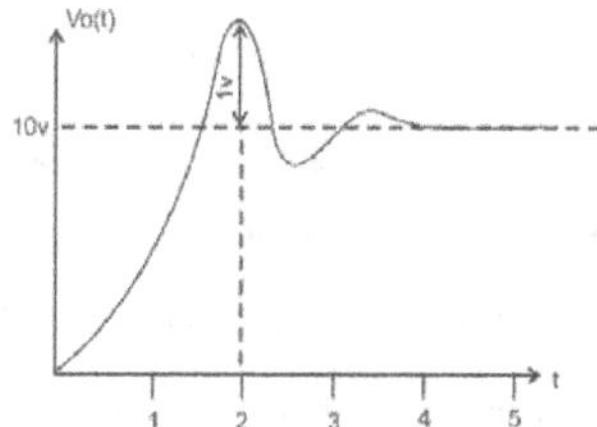

2) In the system shown in figure beside, the numerical values of m, b and k are given as: m= 1 kg, b =2 N-sec/m, and K = 100 N/m. The mass is displaced 0.05 m and released

without initial Velocity. Find the frequency observed in vibration. In addition, find the amplitude four cycles after the system is set into motion. The displacement, x is measured from the equilibrium position.

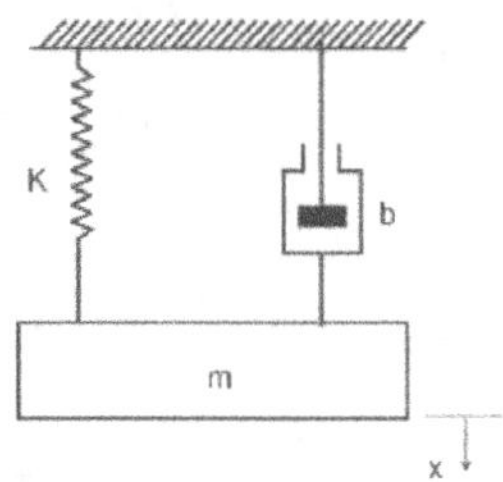

3) What is meant by ***logarithmic decrement?*** Derive an expression for it.

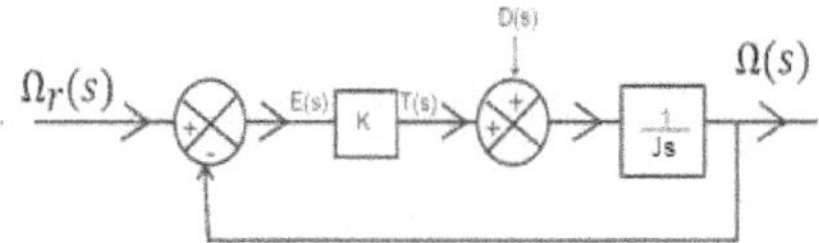

4) The block diagram in figure beside shows a speed control system in which the output member of the system is subject to a torque disturbance. In the diagram $\Omega_r(s), \Omega(s), T(s)$ and $D(s)$ are the Laplace transform of the reference speed, output speed, driving torque, and disturbance torque, respectively. In the absence of a disturbance torque, the output speed is equal to the reference speed. Investigate the response of this system to a unit step disturbance torque. Assume that the reference input $\Omega_r(s) = 0$. In this system, it is desired to eliminate as much as possible the speed errors due to torque disturbance. Is it possible to cancel the effect of a disturbance torque at steady state so that a constant disturbance torque applied to the output member will cause no speed change at steady state?

1.20. Stability

When we want to use/select/design a system, we prefer that the system "behaves well" when it is subjected to inputs and/or disturbances arising out of either users' actions or disturbances from (unpredictable) environment in which the system is used.

The "well behavedness" main mean several things such as reaching desired state very quickly, having no oscillatory dynamics, or that the system does not go "out of control". The last requirement, *viz.*, not going out of control, is central to the discussion in this unit.

We will try to specify in mathematical terms. what it means to say that a system is staying "within control" and then see methods of assessing this property.

Staying "within control", in some sense, means that, when we give a finite input, the output of the system is also finite. this is specified **as Bounded input bounded output(BIBO) Stability.**

BIBO Stability

A system is said to be BIBO stable of its output, y(t), is bounded to bounded input.

Just for the fun of it, we specify the equivalent mathematical expression for BIBO stability of a system with input u(t) and output y(t) of system S. Let M and N be finite positive numbers. Then

$$[S \text{ is } BIBO \text{ stable}] \Leftrightarrow [\{|u(t)| \leq M, \forall t \geq 0\} \Rightarrow \{\exists N \ni |y(t)| \leq N, \forall t \geq 0\}]$$

Carefully observe the notation. The previous statement, written using the symbols, is read as

[S is BIBO stable] _if an only if_ [magnitude of u(t) is less than M _for all_ positive values of t] _implies that_ [_there exists_ a N _such that_ magnitude of y(t) is less than N _for all_ positive values of t]

When we design control system, we intend it to be BIBO stable. Though it is not essential to know the concept of **equilibrium** and **LYAPUNOV stability,** it is instructive to see how mathematical theory involved and how it is connected to the physical world. A casual reader may skip the following material up to **absolute stability and relative stability** without any loss of continuity. We consider a General dynamic system described by first order differential equation of the form

$$x = \underbrace{\frac{dx}{dt} = f(t,x),}_{(T1)} \quad \underbrace{x(t_o) = x_o}_{(T2)} \quad (S)$$

Where $x \epsilon R^n$ is an n- vector called the **state** of the system, $f: R^+ x R^n \rightarrow R^n$ is a vector valued function describing gradient of **state variables**, and $t \epsilon R^+$ is variable to represent **time**. the variable $x_o \epsilon R^n$ specifies the initial values of the state variable. Both (T1) and (T2) together define an ***initial value problem.***

Definition: A Point[11] $x_o \epsilon R^n$ is called an equilibrium point of the system Defined by (S) at time $t_1 \epsilon R^+$ if

$$f(t, x_e) = 0 \quad \forall t \geq t_1$$

Define a r ball of radius r centered at x_e as:

$$B(x_e, r) = \{x \ni |x - x_e| \leq r\}.$$

That is $B(x_e, r)$ is set of all points lying in a sphere of radius r centered at x_e

Definition :- The equilibrium point x_e Of the system (S) is called and isolated equilibrium point if there is an r >0 such that $B(x_e, r) \subset R^n$ contains no equilibrium point of (S) other than x_e . Mathematically we may write:

$$[x_e \text{ is isolated equilibrium}] \Leftrightarrow [\exists r > o \ni \nexists x \in B(x_e, r) \ni \{(f(x_e, t) = 0, \forall t \geq t_1) \wedge (x \neq x_e)\}]$$

Most, if not all, physical systems exhibit isolated equilibrium points. We assume that x_e is an isolated equilibrium point unless stated otherwise. Lyapunov stability theory specifies whether an equilibrium point is stable or not. Without loss of generality $x_e = 0$ is taken to be the equilibrium point.

[Question for the more curious reader: what if $x_e = 0$ is NOT an equilibrium point of the system and we want to use the Lyapunov stability theory?]

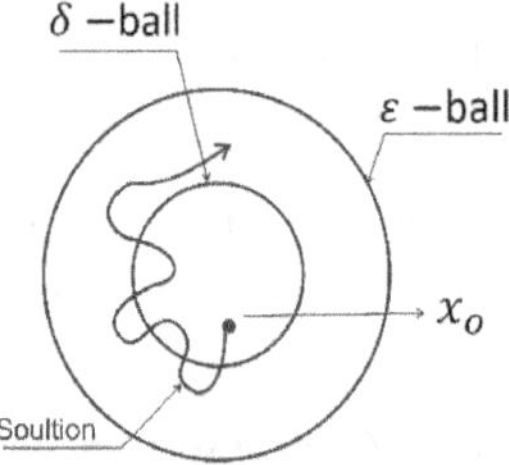

Lyapunov Stability

The equilibrium x = 0 is said to be stable it for every $\varepsilon > 0$, and $t_o \in R^+$ for the system(S), there exist a $\delta(\varepsilon)$ Such that

$$|g(t, x_0)| < \varepsilon \text{ for all } t \geq t_0$$

whenever $|x_o| < \delta(\varepsilon)$, where $g(t, x_o)$ is a solution of (S).

---

[11] R^n is a column vector of n elements where each element is a real number.

In layman's terms, this means, if the equilibrium x_e is stable, then, we can "contain" the solution within a region specified by ε, however small it is, by choosing a δ −ball such that the initial point x_o lies in δ −ball. This stability criterion   is called **stability in the sense of Lyapunov**, abbreviated as (SISL).

In a nutshell, if a system is SISL, its solution is bounded. The next definition, termed **asymptotic stability in the sense of LYAPUNOV (ASISL),** is stricter in the sense that it specifies that the solution <u>converges</u> to equilibrium point, in addition to being contained in a δ −ball.

<u>Definition</u>:- The equilibrium point x = 0 is said to be asymptotic stable if

1. it is stable
2. there exists $\eta > 0$ such that $\lim_{t \to \infty} g(t, x_o) = 0$ wherever $|x_o| \leq n$
    When (ii) is true, the equilibrium x = 0 is called **attractive**.

Thus ASISL specifies condition for attractive equilibrium point x = 0.

The next definition, **exponential stability in the sense of LYAPUNOV (EASISL),** is still stricter condition by specifying how quickly, and in what manner convergence is achieved.

<u>Definition</u>:- the equilibrium x = 0 is exponentially stable if it is asymptotically stable and there exist $\alpha > 0, \beta > 0, \delta > 0$ such that if $\|x(0) - x_e\| < \delta$ then

$$|x(t) - x_e| \leq \alpha e^{-\beta t}|x(0) - x_e| \quad for\ all \ \ t \geq 0$$

Since $\lim_{t \to \infty} \alpha e^{-\beta t} = 0,$ this stability criterion specifies that the solution converges very quickly a long exponential trajectory.

1. Lyapunov stability of an equilibrium means that solutions starting "close enough" to the equilibrium (within a distance δ from it) remain "close enough" forever (within a distance ε from it). Note that this must be true for *any* ε that one may want to choose.
2. Asymptotic stability means that solutions that start close enough not only remain close enough but also eventually converge to the equilibrium.
3. Exponential stability means that solutions not only converge, but in fact converge faster than or at least as fast as a particular known rate proportional to $\alpha e^{-\beta t}$.

For LTI systems, it is known that if it is stable, it is automatically exponentially stable.

Absolute Stability and Relative Stability

If we are merely interested in specifying whether a control system is stable or not, we say that we are talking about ***ABSOLUTE STABILITY***.

On the other hand, knowing that a closed loop control system is stable, it is sometimes possible to characterize the _degree of stability_ or **_degree of instability_**. Such characterization is termed as **RELATIVE STABILITY**. Relative stability involves deciding if one system is **_"more stable"_** or **_"less stable"_** than the other.

It may be noted here in passing that we need not be led to the conclusion that a less stable system is always useless as compared to a more stable system. A notable example of this fact comes from aircraft systems. **_A fighter aircraft is less stable than a commercial Transport aircraft, hence it can maneuver very quickly_**. In fact it can maneuver so quickly that the motion of a fighter aircraft can be quiet violent to the untrained passengers.

There are four methods/criteria that check the stability of LTI system and these are included in the syllabus of the subject.

- <u>ROUTH-HURWITZ CRITERION</u>:- This Criterion is an algebraic method that provides information on absolute stability of an LTI system which has characteristic equation with constant coefficient. Recall that if CLTF is $G(s) = \frac{N_c(s)}{D_c(s)}$, then roots of $D_c = 0$ are called poles and $D_c = 0$ is called characteristic equation. If any of the roots lie in the right hand side of imaginary axis, the system is unstable[12]. RH Criterion also specifies number of roots which have positive Real part. It is also possible to find the range of certain system parameter(s) to ensure absolute stability.

- <u>ROOT LOCUS METHOD</u>:- Root locus method, developed by W.R. Evans, plots the roots of characteristic equation for all values of a system parameter, which is typically system gain. Since this plot is generated from the pole-zero locations of open loop transfer function, effect of adding a pole/zero to OLTF can be visually seen. Though this method also addresses absolute stability, information on relative stability, in **_qualitative sense_**, maybe seen from the way roots migrate as K, system gain varies from 0 to ∞.

- <u>NYQUIST CRITERION</u>:- Nyquist Criterion determine the stability of a closed loop system from its open loop frequency response. This Criterion is a <u>semi graphical method</u> that gives information on the <u>difference between the numbers of poles and zeros</u> of CLTF that are in the right half S plane by observing the behavior of the <u>Nyquist plot</u> of the OLTF. The notions of specifying relative stability quantitatively, in terms of **_Gain Margin_** and **_Phase Margin_**, stem from the Nyquist criteria.

---

[12] Why is this so? Find out about it. I have intentionally left it.

- <u>BODE DIAGRAM</u>:- Bode diagram consists of two frequency response plots: (a) magnitude versus frequency and (b) phase versus frequency. from these two plots very specific information regarding gain margin (GM) and phase margin (PM) or ($\emptyset M$)Is obtained. Using GM and PM, relative stability is quantitatively obtained: ***the more GM and $\emptyset M$, the more is stability of a system.*** Besides this, bode diagram may also be used in **constructing transfer function** of a system from experimental data as we shall see in a later chapter / unit.

The first two methods viz., Routh Huvwitz Criterion and root-locus method fall into time domain analysis design and the next two methods viz. Nyquist Criterion and board diagram fall into frequency domain analyse / design.

Though each method may have its own merits and shortcomings all the methods share one common and ingenious mathematical line of thought: all these topics talk about and provide either qualitative and or quantitative information about closed loop characteristic equation without actually solving it!.

This idea of describing the solution pattern of an equation without solving the equation is further extended to nonlinear Systems by the Russian genius mathematician Lyapunov commenting on the solution of an equation without actually solving it is very radical approach and quizzical: *it is almost akin to knowing the unknown without Actually Knowing It!*

We begin our discussion with Routh-Hurwitz criterion.

Routh-Hurwitz Criterion

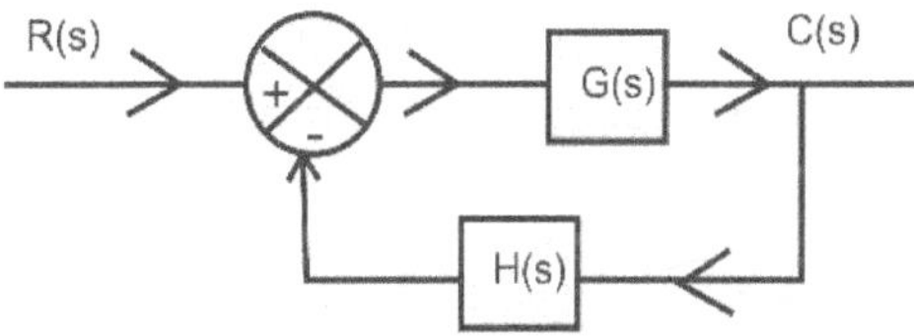

Recall that the closed loop transfer function of the system shown in figure is

$$\frac{C(s)}{R(s)} = \frac{\overbrace{G(s)}^{Closed\ loop\ zeros}}{\underbrace{1 + G(s)H(s)}_{Closed\ loop\ poles}}$$

The characteristic equation is $1 + G(s)H(s) = 0$. The closed loop system is BIBO-stable if all the roots of this characteristic equation have negative real parts.

Rputh-Hurwitz Criterion gives the necessary and sufficient condition for all roots of characteristic equation to lie in the left half of s-plane[13]. The process specified by Hurwitz is slightly tedious and this method was simplified by Routh and hence it is known as Routh-Hurwitz criterion.

The characteristic equation $1 + G(s)H(s) = 0$ may be written in the polynomial form as:

$$a_n s^n + a_{n-1}s^{n-1} + \cdots \ldots + a_1 s + a_o = 0 \quad (A)$$

Where the coefficient are real quantities. We assume that $a_o \neq 0$; that is if the equation has s=0 as a root, it has been removed and we consider equation for remaining roots.

<u>If any one of the coefficient is negative in the presence of at least one positive coefficient, at least one root of (A) has positive Real part and the system is unstable</u>. If all the coefficients of (A) are positive, we construct the Routh's array. in the Routh's array, there are (n + 1) rows when n is the order of equation (A).

<u>ROUTH'S ARRAY</u>:- Construction of Routh's Array is shown below. For each Power of s in the characteristic equation, there is one row. In the left-most column put the powers of s for ease of reference as shown in Table below.

In the first row (s^n), put the alternating coefficients of the terms in characteristic polynomial (A) starting from highest power. In the second row (s^{n-1}), again put the alternating coefficients of the terms in characteristic polynomial (A) starting from the next highest power. The elements in the third row onwards are computed using the previous rows.

For example to compute elements of third row, a_{n-1} is pivot and it appears in the denominator of expressions for all elements in 3rd row, i.e., $b_1, b_2, b_3 \ldots \ldots \ldots$ To memorize, note the "up-going arrow" and "down coming Arrow" in table. Notice that all the arrows begin from elements in first column and go to the 2nd, 3rd, 4th etc columns to find b_1, b_2, b_3 etc. so, we say,

$$b_1 = \frac{a_{n-1}\,a_{n-2} - a_n\,a_{n-3}}{a_{n-1}}, \quad b_2 = \frac{a_{n-1}\,a_{n-4} - a_n\,a_{n-5}}{a_{n-1}}, \ldots \ldots \ldots \ldots$$

$$c_1 = \frac{b_1\,a_{n-3} - a_{n-1}\,b_2}{b_1}, \quad c_2 = \frac{b_1\,a_{n-5} - a_{n-1}\,b_3}{b_1}, \ldots \ldots \ldots \ldots$$

All products arising out of up-going-arrows carry positive sign and all products arising out of down-coming-arrows carry negative sign. Carefully follow through the description and formulae to familiarize yourself with the procedure of generating Routh's array: this will come in very handy.

---

[13] What is this s-plane, BTW.

Table 2: Routh's Array

s^n	a_n	a_{n-2}	a_{n-4}	...	...	...	...	...	...
s^{n-1}	a_{n-1}	a_{n-3}	a_{n-5}	...	...	...	...	...	...
s^{n-2}	b_1	b_2	b_3	...	...	...	...	...	...
s^{n-3}	c_1	c_2	c_3	...	...	...	...	...	...
$\vdots$	$\vdots$	$\vdots$	$\vdots$						
$\vdots$	$\vdots$	$\vdots$	$\vdots$	...	...	...	...	...	...
$\vdots$	$\vdots$	$\vdots$	$\vdots$	...	...	...	...	...	...
s^1	e_1	e_2	e_3	...	...	...	...	...	...
s^0	f_1	f_2	f_3	...	...	...	...	...	...

Alternate coefficients starting from s^n coefficient

Alternate coefficients starting from s^{n-1} coefficient

From third row onwards, elements are computed from previous two rows

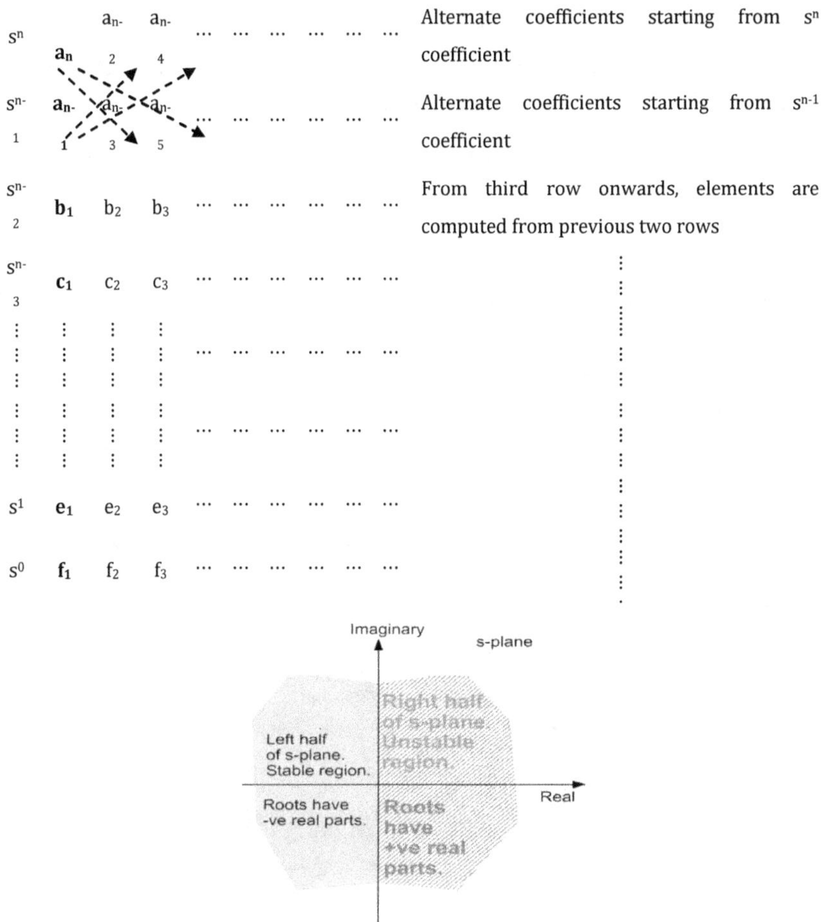

Once the elements are computed, we see the sign of elements in first column, which are shown in red bold: if $a_n, a_{n-1}, b_1, c_1, \ldots\ldots\ldots\ldots, d_1, e_1, f_1$ all have the same sign, then all roots lie in left half of s-plane.

The entire s-plane, plane on which roots are plotted maybe divided into two halves by the imaginary axis: the left half, the stable region, where roots have negative real parts (green shaded, black small letters) and the right half (hatched, red, big letters), the unstable region, where the roots have positive real part.

Routh-Hurwitz Criterion

RH Criterion may be specified as:

The number of sign changes in the first column of Routh's array is equal to the number of roots of characteristic equation which lie in right half of s- plane.

An important question is: should we count the roots lying on the imaginary axis as stable roots lying in left half or as unstable rules lying in the right half ?!? The answer in a way is neither. When CLTF has roots on imaginary axis, the response is oscillatory and we call it sustained oscillations. When such roots exist, the system is <u>marginally stable</u>. Typically a marginal stable system results in when system gain a maximum permissible for stability, K_{max} when system gain $K > K_{max}$, the system becomes unstable. **Also, if** multiple **roots of characteristic equation lie on imaginary axis, the system is unstable**[14].

Example 1: Cubic Equation

The characteristic equation of a system is the cubic equation given below:

$$a_3 s^3 + a_2 s^2 + a_1 s + a_0 = 0$$

Where all coefficients a_3, a_2, a_1, a_0 are positive members.

Determine the condition for stability of the system.

All the elements in column 1 need to be of same sign for stability . Since $a_3, a_2,$ and a_o are positive, the first term in s^1 −row should also be of same sign.

$\therefore a_1 a_2 > a_0 a_3$ is the required condition for stability.

<u>Solution:</u> Construct Routh's array

	Column-1	column-2	column-3
s^3 :	a_3	a_1	0
s^2 :	a_2	a_0	0
s^1 :	$\dfrac{a_2 a_1 - a_3 a_0}{a_2}$	0	
s^0 :	a_0		

Example 2:- RH Criterion and Sustained Oscillations

A closed loop system is shown in figure below. Find the range of gain K for stability of this system.

---

[14] Convince yourself of this before you go any further into this. Read text books, search on web.

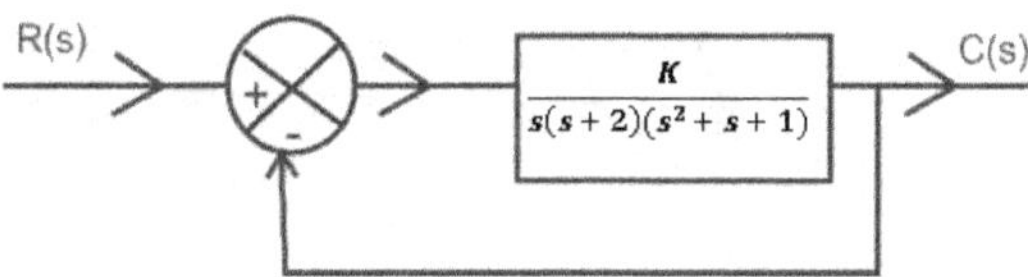

Solution:- Write the characteristic equation:

$$1 + \frac{K}{s(s+2)(s^2+s+1)} = 0$$

$$\boxed{\Rightarrow s^4 + 3s^3 + 3s^2 + 2s + +k = 0} \quad \text{(B)}$$

Construct Routh's array

First Column

s^4 :	1	3	K
s^3 :	3	2	0
s^2 :	$\frac{7}{3}$	K	0
s^1 :	$\frac{14-9K}{7}$	0	
s^0 :	K		

If all the elements need to have the same sign,

$$\frac{14-9K}{7} > 0 \quad \text{and} \quad K > 0$$

∴ The range of values of K for stability is

$$\frac{14}{9} > K > 0$$

Let us further see the problem's implications: $\frac{14}{9} > K \Rightarrow$ system is stable; $\frac{14}{9} < K \Rightarrow$ system is unstable; $\frac{14}{9} = K \Rightarrow$ marginally stable. $K_{max} = \frac{14}{9}$. Substitute this value into characteristic equation and build the Routh's array: $s^4 + 3s^3 + 3s^2 + 2s + \frac{14}{3} = 0$

s^4 :	1	3	$\frac{14}{9}$	
s^3 :	3	2	0	
s^2 :	$\frac{7}{3}$	$\frac{14}{9}$	0	← Auxiliary polynomial $: \frac{7}{3}s^2 + \frac{14}{9} = 0 \Rightarrow s^2 + \frac{2}{3} = 0$
s^1 :	0	0	0	← Row of zeros
s^0 :				

Divide (B) with $\left(s^2 + \frac{2}{3}\right)$ in long division

$$s^2 + \frac{2}{3}\overline{)\ s^4 + 3s^3 + 3s^2 + 2s + \frac{k}{9}(s^3 + 3s + \frac{7}{3}}$$

$$\underline{s^4 + \qquad + \tfrac{2}{3}\,s^2 \qquad\qquad\qquad}$$

$$3s^3 + \frac{7}{3}s^2 + 2s + \frac{14}{9}$$

$$\underline{3s^3 + \qquad + 2s \qquad}$$

$$\frac{7}{3}s^2 + \frac{14}{9}$$

So when $K = \frac{14}{9}$,

$$s^4 + 3s^3 + 3s^2 + 2s + 7 = \left(s^2 + \frac{2}{3}\right)\left(s^2 + 3s + \frac{7}{3}\right) = 0$$

$\therefore s = \pm j\sqrt{\dfrac{2}{3}}$ are two roots that lie on imaginary axis.

When $K = \frac{14}{9}$, the system exhibits sustained oscillations. $\omega = \sqrt{\dfrac{2}{3}}$ rad/sec.

There are two situations which cause complication in completing the Routh's array:

1. when the pivot element is zero
2. when all the element in a row are zero

Example 3: Element in First Column is Zero

As an example of complication (i), consider the system with characteristic equation

$$s^4 + s^3 + 2s^2 + 2s + 3 = 0$$

and build the Routh's array

		When such situation occurs, replace the zero – pivot with a small quantity ε and proceed as shown below:
$s^4 : 1$	2	
3		
$s^3 : 1$	2	$s^4 : 1 \qquad 2 \qquad 3$
0		$s^3 : 1 \qquad 2 \qquad 0$
$s^2 : 0$	3	$s^2 : \varepsilon \qquad 3$
$s^1 :$		$s^1 : \left(2 - \dfrac{3}{\varepsilon}\right) \qquad 0$
$s^0 :$		$s^0 : 3$
Pivot Element = zero		1, 2, and 3 are positive quantities. Hence stability is decided by other quantities, viz. ε & $\left(2 - \dfrac{3}{-\varepsilon}\right)$

If $\varepsilon > 0$, the quantities in first column are:

$$1 \rightarrow +ve$$
$$1 \rightarrow +ve$$
$$\varepsilon \rightarrow +ve$$

$$2 - \frac{3}{\varepsilon} \to -ve$$

$$3 \to +ve$$

In the first column, there is one sign change from positive $\left(\varepsilon \text{ to } \left(2 - \frac{3}{\varepsilon}\right)\right.$ and one more from $\left(2 - \frac{3}{\varepsilon}\right)$ to 3)

If $\varepsilon < 0$, the quantities in first column are:

$$1 \to +ve$$

$$1 \to +ve$$

$$\varepsilon \to -ve$$

$$2 - \frac{3}{\varepsilon} \to +ve$$

$$3 \to +ve$$

In the first column, there is one sign change from (1 to ε and one more from ε to $\left(2 - \frac{3}{\varepsilon}\right)$.

So whether ε is small positive quantity or it is small negative quantity, there are two sign changes and hence 2 roots will be in unstable region of s-plane.

When pivot element is zero, replace it by a small quantity ε and complete the Routh's array.

Example 4: Auxiliary Polynomial

As an example of complication (ii) consider the characteristic equation

$$s^4 + 3s^3 + 5s^2 + 9s + 6 = 0$$

and build the Routh's array

$$
\begin{array}{llll}
s^4 : & 1 & 5 & 6 \\
s^3 : & 3 & 9 & \\
s^2 : & 2 & 6 \leftarrow \text{Auxiliary Polynomial} \\
s^1 : & 0 & 0 \leftarrow \text{Row of zeros} \\
s^0 : & & &
\end{array}
$$

$$A(s) = 2s^2 + 6 \Rightarrow \frac{dA(s)}{ds} = 4s \to 4 \quad 0$$

Explanation

When a row of zeros occurs, the row above is termed Auxiliary polynomial, $A(s)$. In our case, $A(s) = 2s^2 + 6$.

Find $\frac{dA(s)}{ds} = 4$. Substitute [4 0] as elements in place of row of zeroes.

With row of zeros replaced by [4   0], we will continue Routh's Array as:

s^4 :   1   5   6

s^3 :   3   9

s^2 :   2   6

s^1 :   4   0

s^0 :   6

The first column elements do not show a sign change. Hence all roots of the equation lie in stable region.

Thus, in situations where we come across a row of zeros, it is possible to factorize the characteristic equation.

$$2s^2 + 6 \,)\, s^4 + 3s^3 + 5s^2 + 9s + 6 \left(\tfrac{1}{2}s^2 + \tfrac{3}{2}s + 1\right.$$

$$\underline{s^4 + \quad\ \ + 3s^2}$$

$$3s^3 + 2s^2 + 9s + 6$$

$$\underline{3s^3 + \quad\ \ + 9s}$$

$$2s^2 + 6$$

$$\underline{2s^2 + 6}$$

$$\underline{0}$$

$$\therefore s^4 + 3s^3 + 5s^2 + 9s + 6 = (2s^2 + 6)\left(\frac{1}{2}s^2 + \frac{3}{2}s + 1\right)$$
$$= (s^2 + 3)(s^2 + 3s + 2)$$
$$= (s + 1)(s + 2)(s^2 + 3)$$

Thus, in situations where we come across a row of zeros, it is possible to factorize the characteristic equation.

Routh's Criterion may be used to comment about relative stability in a limited way.

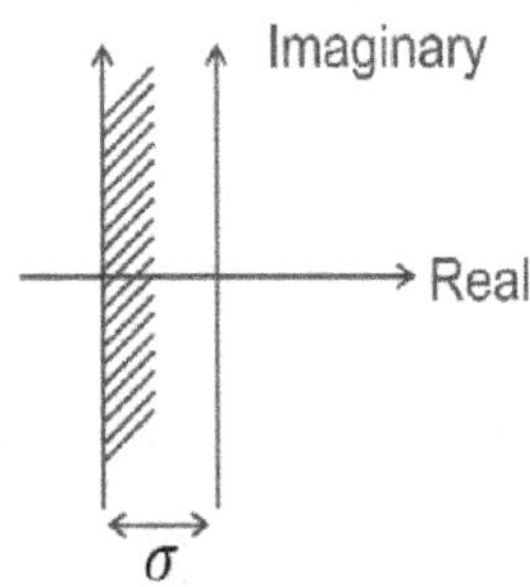

By substituting

$$s = \hat{s} - \sigma,$$

We obtain a polynomial in terms of $\hat{s}$. If we now apply Routh's Criterion to the polynomial, we can find number of roots that are to the right of shifted line.

Thus we make compare how close the dominant poles are to the imaginary axis. Though Routh - Hurwitz Criterion is very handy, it has some limitations:

1. RH Criterion provide specific answers to the question of absolute stability.
2. RH Criterion does not provide an accurate method of dealing with **pure time delay** whereas bode diagram does provide a means of dealing with pure time delay. [What is pure time delay?]
3. RH Criterion does not provide a way of incorporating experimental data into mathematical framework whereas, Bode diagram allows this and thus enables one to provide a way of estimating transfer function.
4. RH Criterion merely specifies whether a system is stable or not. It does not indicate how stability can be improved if needed. Root locus method, Nyquist criterion, and Bode diagram provide information on how stability can be improved.

Home Work 17

1. The characteristic equation of a feedback control system is: $s(s^2 + 3s + 20) + K = 0$. Find out the factored form of the equation when K is such that sustained oscillation occurs for a step input.
2. The feed forward transfer function G(s) for a Unity feedback system is $G(s) = \dfrac{k(s+1)}{s(s+1)(s^2+4s+16)}$

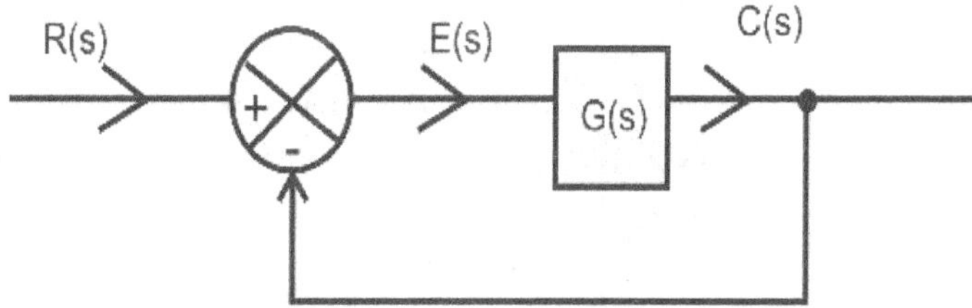

1. Find the range of values of K for stability.
2. Find The frequencies of sustained oscillations.
3. Draw root locus plot for the system.

1.21. Root Locus Method

Introduction

Features of transient response of a closed-loop system are closely related to the locations of closed loop poles, or roots of characteristic equation. Changing the parameters of the system changes the characteristic equation thus causing change in locations of closed loop poles leading to change in features of transient response.

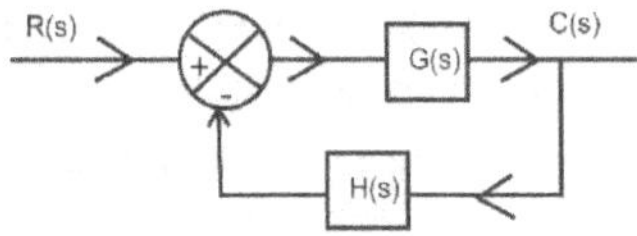

For example, consider the negative feedback system shown in figure beside with the closed loop transfer function

$$\frac{C(s)}{R(s)} = \frac{\overset{zeros\ of\ CLTF}{\overbrace{G(s)}}}{\underset{poles\ of\ CLTF}{\underbrace{1 + G(s)H(s)}}}.$$

The poles of closed loop transfer function are obtained from: $1+G(s)H(s) = 0$ in many cases $G(s)H(s)$ involves a gain parameter K and is of the form

$$G(s)H(s) = \frac{\overset{zeros\ of\ open-loop\ transfer\ function}{\overbrace{K(s + z_1)(s + z_2)\ldots\ldots\ldots(s + z_m)}}}{\underset{poles\ of\ open-loop\ transfer\ function}{\underbrace{(s + p_1)(s + p_2)\ldots\ldots\ldots(s + p_n)}}}$$

We get different roots of characteristic equation for different values of K. Root locus is a plot showing how roots of characteristic equation "move" as K changes its value over all permissible range[15].

When designing a control system, it is required "Push" or "nudge" the closed loop poles to appropriate locations corresponding to desired transient response. Sometimes, it is possible to move the closed loop poles to desired location by adjusting the value of K. In some situations, it may not be possible to move the closed loop poles to desired locations merely by changing gain. in such cases structure of G(s)H(s) may need to be changed by addition of a compensator. in any case, it is useful to know how roots "drift" as K varies from 0 to ∞. The RL method concerns itself with procedure of ***computing and plotting roots of 1+ G(s)H(s) = 0 without actually solving the equation.*** The ***plot of closed-loop poles*** is obtained by noting the

---

[15] What is "permissible range?"

locations of the *poles and zero of open-loop transfer function, G(s)H(s)* and using certain computational procedure.

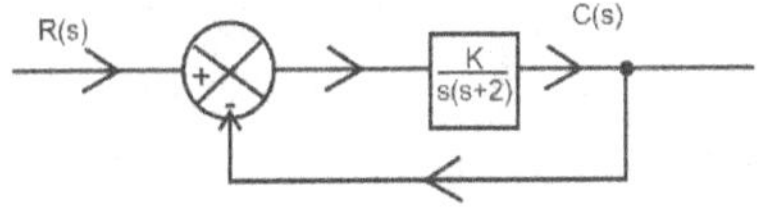

Many a times, modern control engineers tend to think that sketching the root locus by hand is a waste of time since computer packages such as MATLAB generate root locus plot at a mouse-click, so to say. However, experience in sketching the root loci by hand is in valuable for interpreting computer generate root loci as well as for getting a rough idea of the root loci very quickly.

Quick Example

To begin the discussion, consider example of a control system shown in Figure beside whose open loop TF is $\dfrac{K}{\underset{\substack{open \\ loop \\ poles}}{s(s+2)}}$ and whose closed-loop TF is

$$\frac{C(s)}{R(s)} = \frac{K}{s^2 + 2s + K}.$$

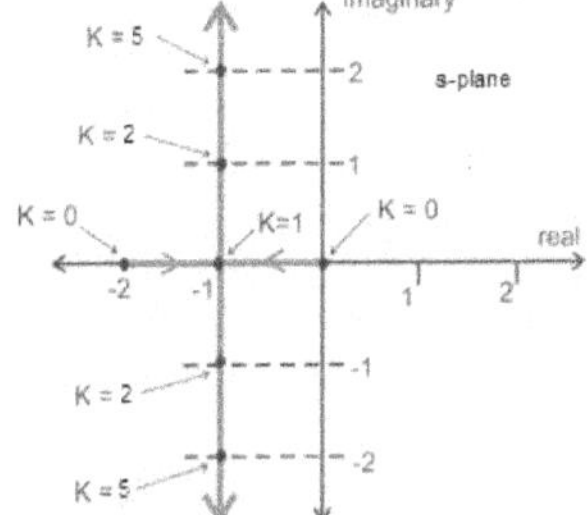

The characteristic equation is $s^2 + 2s + K = 0$. Look at table which shows roots for various values of K. Since the characteristic equation is quadratic, for each value K, there are two roots. We can put the values of these roots on a graph sheet as shown in s-plane.

K	Roots
0	0, -2
1	-1, -1
2	-1 ± J
5	-1 ± 2J

Notice that when K = 0, the roots[16] are 0 and -2. As K is gradually increased, these roots move close to each other and they "collide" at -1. From then on, increasing K will make the roots go vertically upward. When the path of roots is connected as shown in red, thick line, it is called root-locus plot. Since the characteristic equation is a quadratic equation, it was possible to draw root locus very easily. If the order of characteristic equation is more than 3, it is very difficult to draw root locus this way. In what follows, a simplified method, developed by W.R. Evans will be presented.

First, a few preliminaries will be presented regarding the plot drawn in the previous page. Because the roots of characteristic equation may be real, imaginary, or complex, they are shown on a graph sheet with real, imaginary axes shown on it. Such a graph sheet (well, any sheet on which these are plotted) is called **s-plane**. s-plane is divided into (i) left-half where the real-parts of roots are negative and (ii)right-half where the real-parts of roots are positive.

The roots begin their "journey" when K=0 at s = 0 and s = - 2. s = 0 and s = - 2 happen to be the poles of $\frac{k}{s(s+2)}$. This is not coincidence. In all root loci, the journey of roots begin at poles of feed-forward transfer function. Notice that we didn't have to solve characteristic equation to identify the beginning point. There will be as many branches in the root locus as there are poles in feed forward transfer function and each branch begins at a pole, and depending on the number of zeros, they either terminate at a zero or continue to going to Infinity along asymptotes.

Also, it may be noticed that in example considered, one branch begins at s = 0 and begins "journey" to the left. Another branch begins at s = - 2 and begins its journey to the right. these two "collide" at s = -1 and "break-away" from real axis. The point s = -1 is termed "break-away point".

Leaving all mathematical details, $\frac{dK}{ds} = 0$ gives breakaway or break in points. Further, at each breakaway or break in point characteristic equation has at least a pair of equal[17] roots. This may be compared to critical damping.

The two branches go to infinity: one branch contain roots with positive Imaginary part and another with negative imaginary part.

To express this face, we say that there are two asymptotes: one at 90° and other at 270 °. These two are called **angles of asymptotes**. further, the asymptotes meet at s = -1.

---

[16] It is not a coincidence that the plot of closed-loop poles begins at the open-loop poles. It always does.
[17] See what is the value of K at break-away point in previous example. For this value of K, what are the roots?

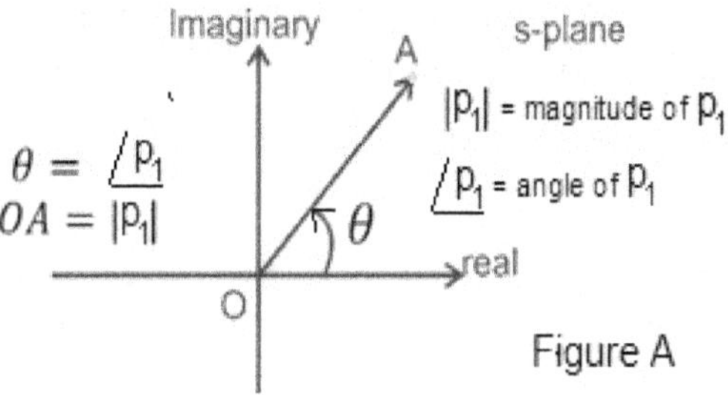

The example we considered has only two feed forward-poles and no zeros. But the line of thought we followed for the simple case can be used even when the feed forward transfer function has more poles and/or zeros.

All the information for plotting the root-locus may be obtained from feed forward transfer function and there is no need to compute the roots of characteristic equation. In what follows, an logarithm for plotting root locus will be presented without giving any mathematical details

Before we delve into those details, let us note that all points on s-plane have real part and imaginary part just as they have a magnitude and an angle. We say that "s" is a point in the complex plane, for example, the point A.

The magnitude of "s" is the length of the line, OA, joining origin and A. The angle of "s" is the angle[18] OA makes with positive real axis.

Rules for Plotting Root-Locus

The rules for plotting root-locus will be presented without mathematical proof and the magnitude-angle concept is invoked wherever need is felt.

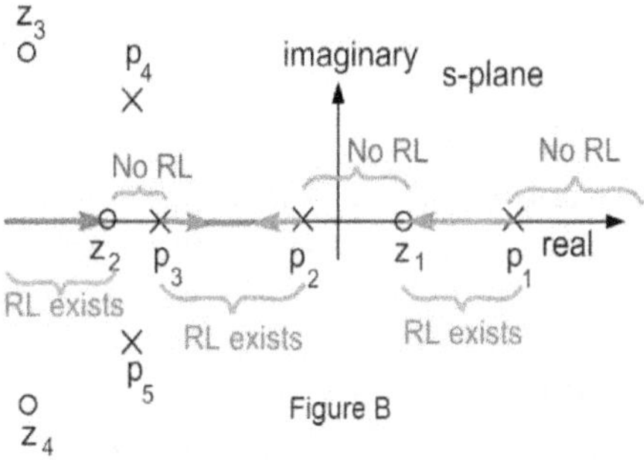

---

[18] Notice that we measure counter-clock-wise angles as positive.

Plot the poles and zeros on the s-plane. Each Pole is shown with a symbol "x" and each zero is shown with a symbol "o". Let n = number of poles of G(s)H(s) And let m=number of zeros of G(s)H(s). Let $p_1, p_2, \ldots \ldots p_n$ be the poles and let $z_1, z_2, \ldots \ldots, z_m$ be the zeros.

1. RL on Real Axis

If some poles and/or zeros lie on real axis, portions of root locus plot lie real axis. To identify the segments of RL on real axis use following discussion. In the discussion, the word "entity" is used to mean either pole or zero.

- Towards the right hand side of right most entity (p_1 in Figure B),   root locus do not exist.
- Between the rightmost pair of entities (p_1 and in z_1 Figure A), RL exists. Darken the part of real axis between rightmost pair of entities.
- Between the next rightmost pair of entities (z_1 and p_2 Figure B), RL doesn't exist.
- Between the next right most pair(p_2 and p_3 in Figure B), RL exists and so an alternately
- *NOTE THAT COMPLEX CONGUGATE POLES (p_4 and p_5 in Figure A)/ ZEROS($z3$ and z_4 in Figure B) DO NOT AFFECT RL ON REAL AXIS.*

When we say that RL exists in a given region, we mean that every point in the region lies on the root-locus. Also observe carefully how arrowheads are assigned: at each pole, the branch goes away and at each zero branch comes in.

2. Asymptotes

If $n \neq m$, there will be (n-m) asymptoes. These asymptoes meet at a point, termed **centroid**, on RL, given by σ_1. The formulae for computing coordinate centroid lying on real-axis, σ_1 and the angles between the asymptotes and real axis, $\theta_i, i = 1, \ldots \ldots \ldots (n - m)$, are:

$$\sigma_1 = \frac{\left(\sum_{i=1}^{n} p_i\right) - \left(\sum_{j=1}^{m} z_j\right)}{(n\text{-}m)} \qquad \theta_i = \frac{(2i + 1)}{(n - m)} \times 180^{\circ}$$
$$i = 1, 2, 3, \cdots, (n - m)$$

Show centroid on the real axis and draw straight lines from the centroid to represent asymptotes at angles θ_i as construction lines. Branches of root-locus go to infinity along these lines.

3. Break-in or Break-Away Points

If there are poles/zeros which are complex conjugate in addition to those on real axis, then maybe <u>break away or in points</u>, given by equation, on the real axis:

$\frac{dK}{ds} = 0$. Compute the roots of this equation and show on s-plane.

4. Intersection with Imaginary Axis

Identify points of intersection of RL with imaginary axis using Routh-Hurwitz criterion. Notice that if asymptotes intersect the imaginary axis, branches of root-locus also intersect imaginary axis. Typically this situation occurs when K is at its maximum value for stability.

5. Angle of Arrival or Departure

If there are complex conjugate poles, **angle of departure** needs to be computed and if there are complex conjugate zeros, **angle of arrival** needs to be computed.

Let p and $\bar{p}$ be the complex conjugate poles at which the angle of departure needs to be computed. Then, the angle of departure is given by,

$$\varphi_d = 180\,° - \left[\left(\sum_{i=1}^{n} \angle(p - p_i) \right) - \left(\sum_{j=1}^{m} \angle(p - z_j) \right) \right]. \quad (A)$$

In equation (A), $\angle(p - p_i)$ is the angle made by the complex number $s = (p - p_i)$ with real axis as explained in Figure A. Similarly, $\angle(p - z_j)$ is the angle made by the complex number $s = (p - z_j)$.

If we draw a tangent to the root-locus at the complex pole p, this tangent will make an angle φ_d with the real-axis. The angle of tangent to the root-locus at $\bar{p}$ will be $-\varphi_d$.

Similarly, the angle of arrival, φ_a, at the complex pole z is computed as and the angle of arrival at the pole $\bar{z}$ is noted to be $-\varphi_a$.

$$\varphi_a = 180\,° - \left[\left(\sum_{i=1}^{n} \angle(z - p_i) \right) - \left(\sum_{j=1}^{m} \angle(z - z_j) \right) \right]. \quad (C)$$

The angles φ_a and $-\varphi_a$ are the angles made by the tangents to the root-locus drawn at z and $\bar{z}$ respectively.

The computation of angle of departure/arrival is based on the **angle criterion** discussed at the end of this section.

6. K at Salient Points

After plotting the root-locus, the value of K at each of the salient points (break-in, break-away, intersection with imaginary axis) needs to be shown. Let **s be a salient point on the root-locus** where the value of K needs to be computed. Then

$$K = \frac{|(s - p_1)| \cdot |(s - p_2)| \cdot |(s - p_3)| \cdot \cdots \cdot |(s - p_n)|}{|(s - z_1)| \cdot |(s - z_2)| \cdot |(s - z_3)| \cdot \cdots \cdot |(s - z_m)|}. \quad \text{(B)}$$

The value of K at a point **s** *lying on the root-locus* is computed using **magnitude criterion** discussed at the end of this chapted.

Following this procedure takes a bit of practice. Solve at least 4 problems and the process will be very clear.

Example 1: System with Only Poles

Sketch the root locus plot for the system shown in figure below as ke varies from 0 to ∞.

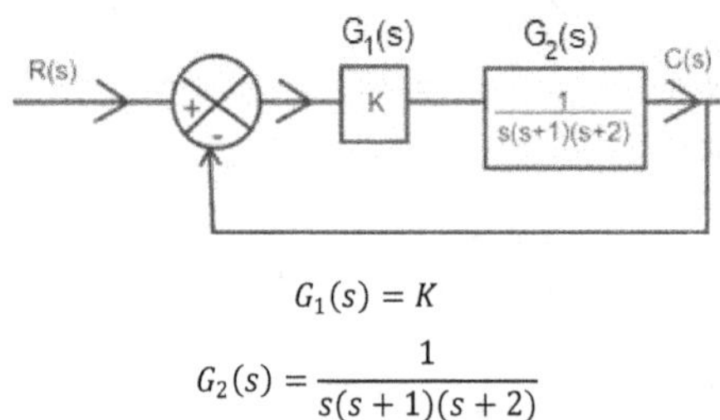

$$G_1(s) = K$$

$$G_2(s) = \frac{1}{s(s + 1)(s + 2)}$$

Solution

The open loop transfer function, $G(s) = G_1(s)G_2(s) = \frac{K}{s(s+1)(s+2)}$

<u>Step 1</u>:- identify poles and zeros of OLTF. There are three poles, p_1, p_2, and p_3 so n=3. There are no zeros so m=0.

$$p_1 = 0$$
$$p_2 = -1$$
$$p_3 = -2$$

Hence, there are n - m = 3 asymptotes. The poles and zeros along with the root-locus plot are shown on Graph Sheet 1.

<u>RL on real axis</u>:- Draw RL on real axis: towards RHS of p_1, no root locus exists. Between p_1 and p_2 RL exists. Between p_2 and p_3 RL does not exist. To the left of p_2 RL exists. Draw straight lines in locations where RL exists. Branch of root-locus starting from p₁ will go to the left and the branch of root-locus starting from p₂ will go to the right. Show the directions of these branches. The branch of root locus starting from p₃ will move to the left. This branch goes along the asymptote at 180° to infinity.

<u>Asymptotes</u>:- Since $n \neq m$, there are $n - m = 3$ asymptotes. These asymptotes meet on real axis at σ given by:

$$\sigma = \frac{\sum_{l=1}^{p} pi - \sum_{l=1}^{z} zi}{n-m} = \frac{(-1-2)-(0)}{3} = -1.$$

The angles of these asymptotes are: $\theta_i = \frac{180*(2i+1)}{n-m}$, $i = 1, 2, 3$

$$\theta_1 = \frac{180*(2+1)}{3} = 180°; \ \theta_2 = \frac{180*(4+1)}{3} = 300; \ \theta_3 = \frac{180*(6+1)}{3} = 420°$$

In terms of positive and negative angles, these will be, 180°, -60°, +60°. Show these asymptotes on graph sheet. The 180° asymptoes is negative real axis itself. Hence the other two are shown. The asymptoes are shown light since these are construction lines.

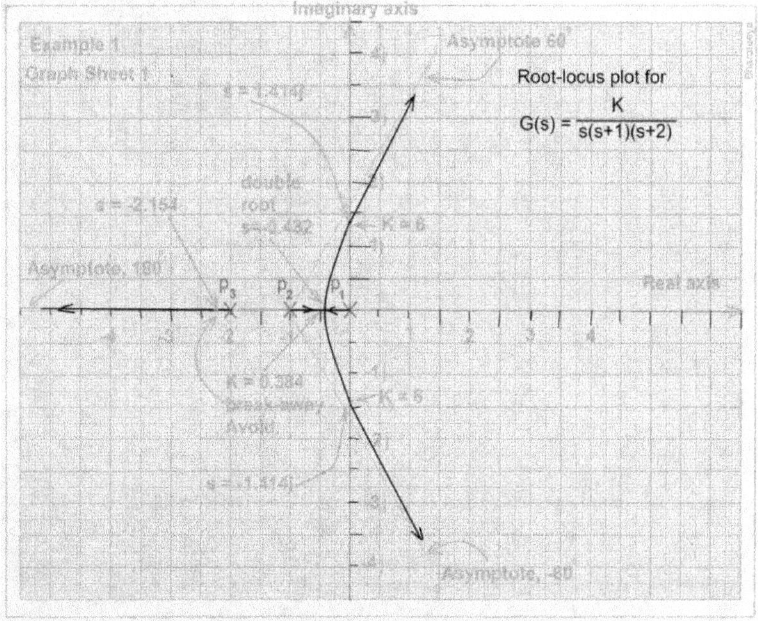

Break-away points:- One branch of RL starts at p_1 and and moves to the left. One branch of RL starts at p_2 and moves to the right. These two branches collide and get separated from real axis at break away point. To find the break away point, We write the characteristic equation:

$$1 + G(s)H(s) = 0 \Rightarrow 1 + \frac{K}{s(s+1)(s+2)} = 0$$

$$\Rightarrow s(s+1)(s+2) + K = 0$$

$$\Rightarrow K = -s(s+1)(s+2)$$

And write the equation

$$\frac{dK}{ds} = 0$$

to get

$$3s^2 + 6s + 2 = 0.$$

$$\Rightarrow s = \frac{-6 \pm \sqrt{36 - 21}}{6} = \underbrace{-0.432}_{We\ accept\ this.} , \quad \overset{We\ reject\ this.Why?}{-1.577}$$

Show break – away point s =- 0.432 on graph sheet.

<u>Intersection with imaginary axis</u>:- to identify point of intersection of RL with imaginary axis, recall the characteristic equation

$$s(s + 1)(s + 2) + K = 0 \Rightarrow s^3 + 3s^2 + 2s + K = 0$$

And build Routh's array

$$s^3 : 1 \qquad 2 \qquad \text{when } K > 0, 6 - K > 0, \text{ system is stable}$$

$$s^2 : 3 \qquad K \qquad \text{when K=6, Marginal stable}$$

$$s^1 : \frac{6-K}{3} \qquad 0 \qquad \therefore \text{ one pair of roots exist on imaginary axis}$$

$$s^0 : K \qquad \qquad 3s^2 + 6 = 0 \Rightarrow s^2 + 2 = 0$$

$$\Rightarrow s = \pm j\sqrt{2} \ \text{ are points on RL.}$$

Sow these points on graph sheet.

{Angle of departure/arrival need not to be computed since there are no complex conjugate poles or zeros.}

<u>Draw RL Plot</u>: Join the points $s = -0.422, j\sqrt{2}, -j\sqrt{2}$ and draw freehand curve and extend it so that it gets tangential to the asymptoes at large value of $|s|$ or $K \to \infty$.

Compute the value of K at $s = -0.432$ $\ as \ \ K = |-s(s + 2)(s + 1)|_{s=-0.432}$

Or $K = |-0.432 * (-0.432 + 1) * (-0.432 + 2)| = 0.384.$

Show this and values of K at $\pm j\sqrt{2}$ to complete the root locus plot.

It is to be noted that K = 0.384 or values of K very near to 0.384 needs to be avoided. Think why? Similarly K = 6 or value of K very near to 6 needs to be avoided.

Example 2: System with Complex Conjugate Poles

Sketch root locus plot for the system shown in figure below for K varying from 0 to ∞ and H(s) = 1.

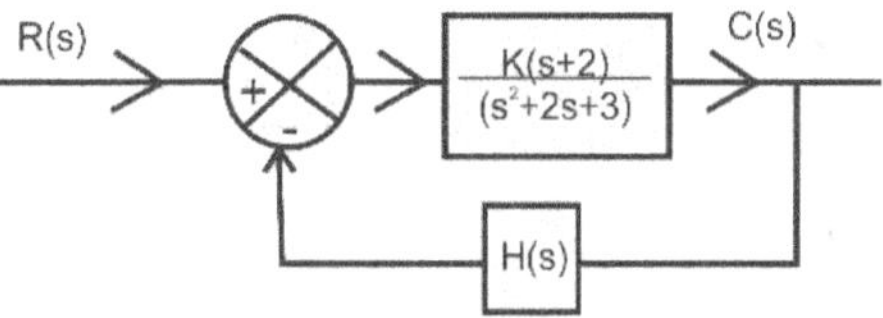

Solution

Identify poles and zeros:- $p_1 = -1 + j\sqrt{2}, \ p_2 = -1 - j\sqrt{2} \qquad \therefore n = 2$

$$z_1 = -2 \qquad\qquad \therefore m = 1$$

Poles and zeros and the root-locus plot are shown on Graph Sheet 2.

<u>Step 1</u>: It can be seen very clearly that RL exists only to the left of the zero z_1 on real axis.

<u>Step 2</u>: There is only one asymptote since $n - m = 1$. $\sigma = \frac{p_1 + p_2 - z_1}{1} = 0$ & $\quad \angle s = 180^\circ$.

The asymptote goes along negative real axis and passing through origni.

<u>Step 3</u>:- There will be one break-in point which is obtained by writing characteristic equation:

$$(s^2 + 2s + 3) + K(s + 2) = 0 \Rightarrow K = -\frac{s^2 + 2s + 3}{s + 2}$$

and setting

$$\frac{dK}{ds} = 0$$

$$\therefore (s + 2)(2s + 2) - (s^2 + 2s + 3)(1) = 0$$

$$\Rightarrow s^2 + 4s + 1 = 0$$

$$\Rightarrow s_1 = -0.268, -3.732$$

<u>Step 4</u>:- To obtain points of intersection with imaginary axis, build Routh's Array.

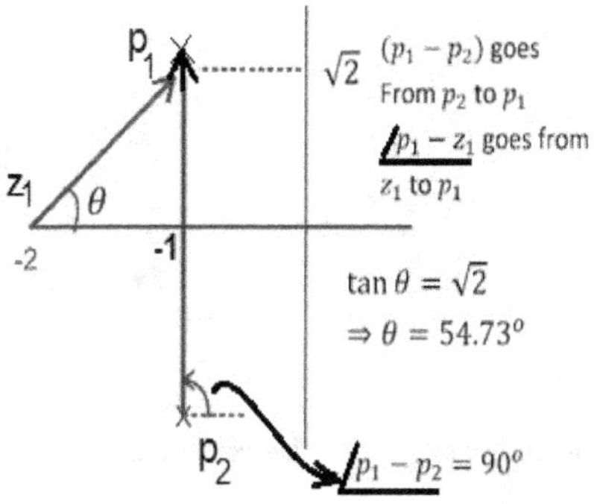

$$s^2 : 1 \qquad (3 + 2K)$$
$$s^2 : (2 + K) \qquad 0$$
$$s^0 : (3 + 2K)$$

There are no sign changes in first column.

$\therefore$ No points of inter section with imaginary axis and system is stable $\forall K > 0$.

<u>Step 5</u>:- To find angle of departure at p_1, use equation (A).

$$\varphi_{d1} = 180° - \{\angle(p_1 - p_2) - \angle(p_1 - z_2)\}$$
$$\Rightarrow \varphi_{d1} = 180 - \{90 - 54.73\} = 144.73 .$$

φ_{d1} is the angle between the tangent to the root-locus at p_1 and the real axis. Similarly, $\varphi_{d2} = -\varphi_{d1}$ is the angle between the tangent to the root-locus at p_2 and the real axis.

<u>step 7</u>:- Value of K at break-in point: $K = \left| \dfrac{s^2 + 2s + 3}{s + 2} \right|_{s = -3.732} = 5.464$

Draw RL such that it passes through; $p_1, s = -3.732, p_2$. This curve needs to be tangential to lines drawn at p_1 and p_2 making angles φ_{d1} and φ_{d2} respectively.

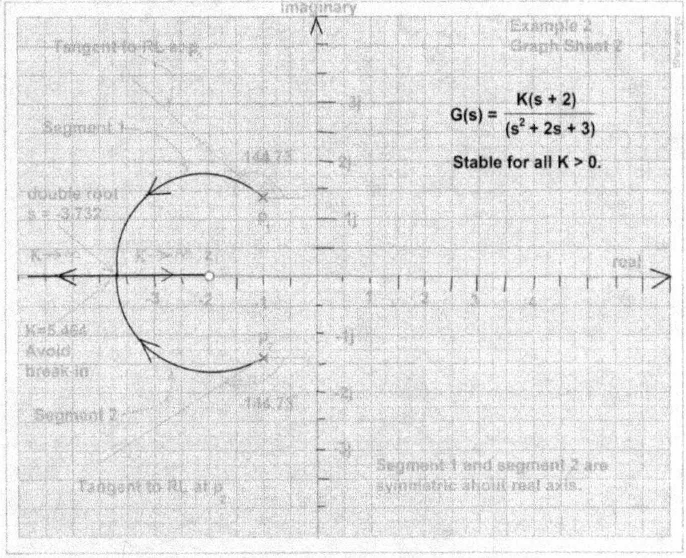

s- Plane Angle Criterion and Magnitude Criterion

Since roots can be complex quantities, they are represented on a sheet with real/ imaginary axis, and this shit is termed "s-plane".

Consider complex conjugate poles P_1 and P_2 shown by "x" marks and z_1, z_2, complex conjugate zero shown by "o" marks in s- plane shown besides. the complex number P_1 may be represented by vector going from origin O to A. $|||^{by}$ P_2 may be represented by directed line segment OB. Similarly we say $\overrightarrow{OC} = z_1$ and $\overrightarrow{OD} = z_2$. Consider a point and as shown in s- plane. Carefully note that by Triangle Law of addition, the directed line segment going from z_1 to s is the vector quantity $(s - z_1)$. Similarly that Line segment growing from p_1 to z_1 is $(z_1 - p_1)$.

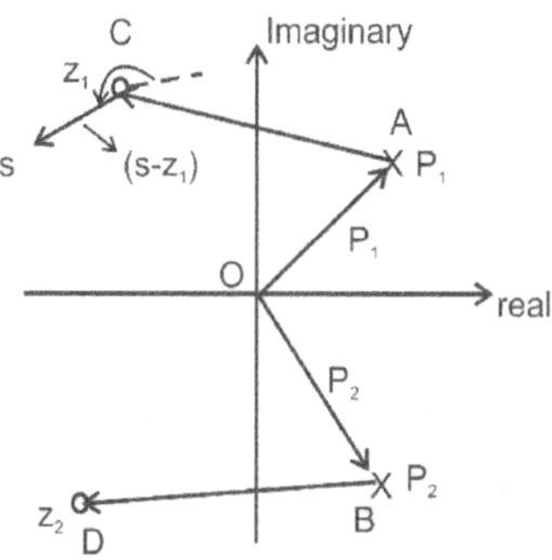

Understand the previous paragraph thoroughly by building your own examples and then only go to subsequent discussion.

Now consider an open loop transfer function of the form shown below. Since s is complex, GH(s) is also complex number. Further, we assume n ≥ m and that K is positive real number.

$$G(s)H(s) = \frac{K(s - z_1)(s - z_2)\ldots\ldots(s - z_m)}{(s - p_1)(s - p_2)\ldots\ldots(s - p_n)}.$$

The characteristic equation for the closed loop system is

$$1 + G(s)H(s) = 0$$

$$\Rightarrow (s - p_1)(s - p_2)\ldots\ldots(s - p_n) + K(s - z_1)(s - z_2)\ldots\ldots(s - z_m) = 0$$

$$\Rightarrow -K = +\frac{(s - p_1)(s - p_2)\ldots\ldots(s - p_n)}{(s - z_1)(s - z_2)\ldots\ldots(s - z_m)} \quad (D)$$

Since $p_1, \ldots\ldots, p_n, z_1, z_2, \ldots\ldots z_m$ and s can be complex, the fraction on the right hand side of equation may also be complex. Note that in this equation, $(s - p_1)$ is a complex vector going from p_1 to s and $(s - z_1)$ is a vector going from z_1 to s and so on. Since, (D) involves complex quantities, magnitudes must be same on either side. Hence,

$$|-K| = K = \frac{|(s - p_1)| * |(s - p_2)| * |(s - p_3)|\ldots\ldots(s - p_n)}{|(s - z_1)| * |(s - z_2)| * |(s - z_3)|\ldots\ldots(s - z_m)} \quad (E)$$

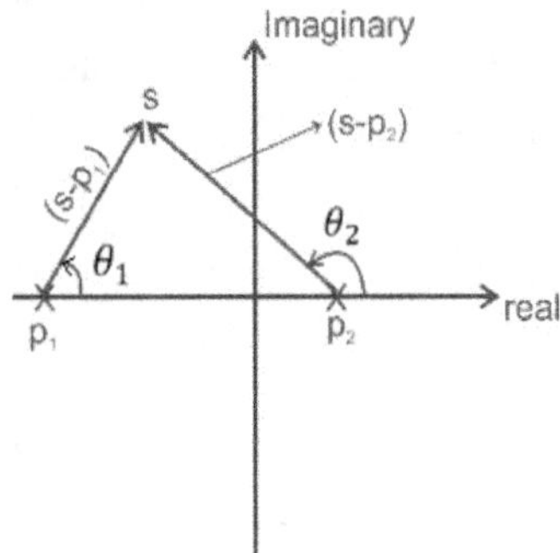

Equation (E) is called Magnitude criterion. Magnitude criterion is used to find the value of gain K at a point lying on Root Locus.

Since, quantities on either side of equation (E) are complex, the angles on either side must also be equal. Recall that angle of a vector, $\theta_1 = \angle(s - p_1)$, is defined as the angle made by line going from p_1 to s with positive real axis.

We write: $\theta_1 = \angle(s - p_1)$

We also know that $\angle(s - p_1) * (s - p_2) = \angle(s - p_1) + \angle(s - p_2) = \theta_1 + \theta_2$

And $\angle\left(\frac{(s-p_2)}{(s-p_1)}\right) = \angle(s - p_2) - \angle(s - p_1) = \theta_2 - \theta_1$

With this notation, definitions, and mathematical equations, angles equated on either sides of equations (D) yields

$$[\angle(s - p_1) + \cdots + \angle(s - p_n)] - [\angle(s - z_1) + \cdots \ldots + \angle(s - z_m)] = \angle - K = 180°\pm 360i$$

$$\Rightarrow \left[\sum_{i=1}^{n} \angle(s - p_i)\right] - \left[\sum_{j=1}^{m} \angle(s - z_j)\right] = 180°\pm 360i \qquad \text{(F)}$$

Equations (F) is called Angle criterion. The RL can be constructed from this angle criterion alone. Equation (F) can be used to find if a point "s" lies on the RL of a system with open loop poles $p_1, \ldots \ldots p_n$, and $z_1, \ldots \ldots, z_m$.

Home Work 18

1) If $s_1 = -3 - 4j$ and $s_2 = 3 + 4j$ where $j = \sqrt{-1}$, find

$|s_1|, |s_2|, |s_1 + s_2|, \left|\frac{s_1}{s_2}\right|, |s_1 s_2|, \angle s_1, \angle s_2, \angle(s_1 + s_2), \angle\left(\frac{s_1}{s_2}\right), \angle(s_1 s_2).$

2) An open-loop transfer has the pole-zero configuration as shown in figure below. If the gain parameter is K,

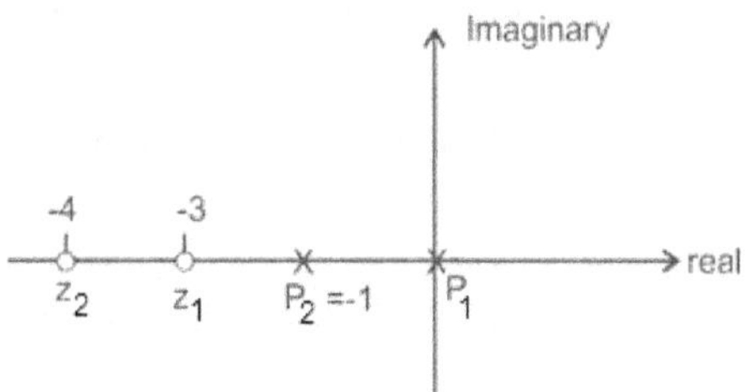

1. Find the open loop TF
2. Without sketching the Root Locus, is it possible to decide if s = 3+6j lies on root locus?

3) For a system, the open-loop Transfer Function is $G(s) = \dfrac{K(s+1)}{s(s-1)(s^2+4s+16)}$
   Find $G(s), |G(s)|$, and $\angle G(s)$ when K = 10 and $s = 1+j$.

4) Draw root locus plot for a system whose OLTF is $G(s) = \dfrac{K}{s(s+4)(s+5)}$ as the gain K, a positive quantity varies from 0 to ∞ Show all details.

5) Draw root locus plot for a system whose OLTF is $G(s) = \dfrac{K(s+1)}{s(s+4)(s+5)}$ as K varies from 0 to ∞. Show all details.

6) Draw root locus plot for the system with OLTF G(s) given in (Q.3). Solve this problem for sure.

1.22. Frequency Domain Analysis

Performance of control system is often specified based on time responses to test signals. However, as discussed in previous chapter, the time response of a control system is more difficult to determine analytically, especially in case of systems with order three or more. because of this and the fact that communication systems deal with frequency response since most of the signal to be processed or either sinusoidal are composed of sinusoidal components. the following reasons form the motivation to study frequency response of control system:

a) Communication system analysis and design involved dealing with signals that are either sinusoidal or composed of sinusoidal components of various frequencies.

b) Time response of a control system is usually more difficult to determine analytically, especially for higher order systems.

c) There are no standardized method for designing a control system to meet time domain performance specifications such as rise time, peak time, peak overshoot, settling time, and so on.

d) In the frequency domain, there is a wealth of graphical methods available which are not limited to lower order systems

e) It is more convenient for measurements of system sensitivity to noise and parameter variation.

f) Relative stability for LTI systems is very precisely specified using gain margin and phase margin, concepts stemming from frequency response.

g) There are correlating relations between time domain and frequency domain. Hence, that time domain characteristics can be "predicted" based on the frequency domain characteristics.

h) Transfer function can be estimated from the experimental analysis on "unknown" systems from Bode diagram.

i) System with pure time delay can be analyzed without resorting to approximation.

Besides these reasons, frequency response provide an alternative view point for control system problems involving analysis and design.

To begin the study of frequency domain analysis of a linear system, recall that when the input to a linear time invariant system is sinusoidal, the output is also a sinusoid with same frequency but possibly different amplitude and phase.

Let $r(t) = R\,Sin(wt)$ be the input of a system with transfer function $M(s)$. then, the steady-state output, $y(t)$ may be specified as: $y(t) = Y\sin(\omega t + \emptyset)$ where Y is the amplitude of the output since wave and $\emptyset$ is phase shift. The ratio $\frac{Y}{R}$ and the angle $\emptyset$ constitutes the basic of all the frequency-domain analysis we study. Knowing the transfer function $M(s)$ of the system, $\frac{Y}{R}$ and $\emptyset$ can be deduced.

Let us consider the closed loop system shown in the figure below

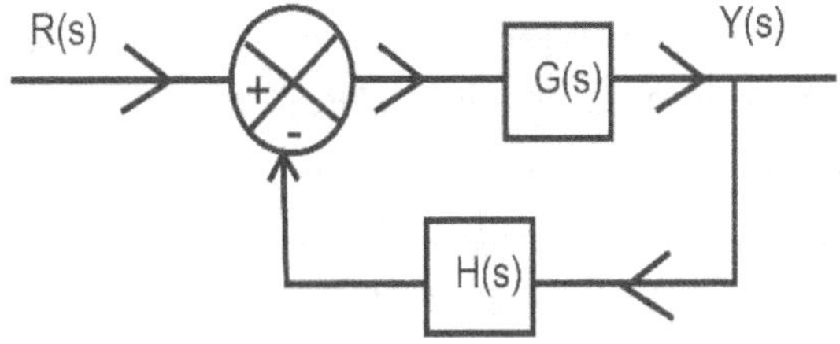

for which the transfer function is obtained as:

$$M(s) = \frac{Y(s)}{R(s)} = \frac{G(s)}{\underbrace{1 + G(s)H(s)}_{\text{defined as } \Delta(s)}} \quad (A)$$

Thus, for an input of $R(s)$, $Y(s) = M(s)R(s)$. The beauty of frequency-domain analysis is that simply by substituting $j\omega$ in place of s in $M(s)$, the magnitude and phase information is obtained:

$$M = \frac{Y}{R} = \text{steady} - \text{state amplitude ration} = |M(j\omega)|$$

$$\emptyset = \text{steady} - \text{state phase} - \text{shift} = \angle M(j\omega)$$

The two quantities can be used in graphical method in one of the two ways:

Polar Diagram

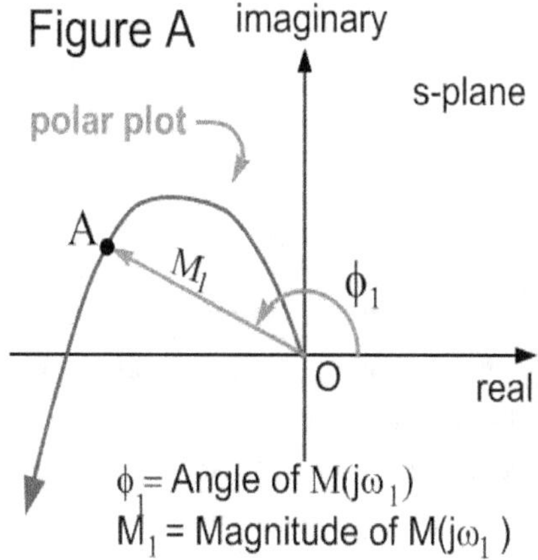

For each value of ω in the range $[\omega_1, \omega_2, \ldots \ldots \omega_n]$, the amplitude ratio $[M_1, M_2, \ldots \ldots M_n]$ and the phase value $[\emptyset_1, \emptyset_2, \ldots \ldots \emptyset_n]$ are computed and plotted as a polar plot shown. For example, a point A is located by drawing a line making an angle $= \emptyset_1$ and on that line, measuring a length M_1.

Thus, in Figure A, the length OA $= M_1$. Notice that, in locating point A, both the magnitude information and phase information are shown on plot but the frequency information is not shown on the plot.

Thus, frequency information is not directly visible on the polar plot.

Bode Diagram

This diagram consists of two plots drawn with a common $\omega -$axis. Figure below shows an example.

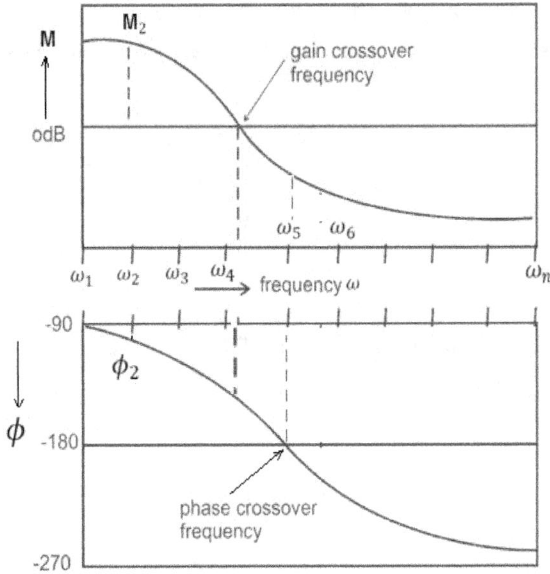

In the magnitude plot, the vertical axis, M, is measured in decibels i.e., vertical axis is $20\ log_{10} M$. In the phase plot, $\emptyset$ on vertical axis is shown in degrees. Both these plots share the common frequencies $\omega_1, \omega_2, \ldots \ldots \omega_n$. These frequency values are chosen in such a way that both gain crossover and phase crossover are captured. To enable us to cover wide range of ω −values, frequency is plotted on logarithmic scale and $M_{dB}, \emptyset$ are shown in linear scale. Semi-log sheets are used to draw the M_{dB} Vs ω plot and $\emptyset$ Vs ω plot and typically both these are drawn on the same sheet. Very many times, when it is understood from the context, the "dB" in subscript is omitted.

Frequency Domain Specifications

Similar to the time-domain specifications, there are specifications in the frequency domain. These specifications enable us to compare the performance of one system with that of the other. There are four specifications which are listed below.

1. Resonant Peak, Mᵣ

Resonant Peak is the maximum value of $|M(j\omega)|$. The desired value of M_r is chosen between 1.1 and 1.5. Large value of M_r corresponds to a large peak overshoot to step response. In general M_r gives an indication on the relative stability: very large Mr indicates very less damping and less stability.

2. Resonant Frequency, ω_r

The resonant frequency, ω_r, is the frequency at which the peak- resonance M_r occurs. It is often desirable to have ω_r much above the operating speeds.

3. Band Width BW

Bandwidth in some sense defines the operating frequency range. It is defined as *the frequency at which $|M(j\omega)|$ drops to 70.7% of, or 3dB down from, its zero frequency value.*

Large BW corresponds to faster rise time and higher frequency signals are more easily passed through the system. BW is also indicative of noise filtering characteristics and the robustness of the system. The robustness represent a measure of the sensitivity of a system to parameter variations. A robust system is one that is insensitive to parameter variations.

4. Cut Off Rate

The slope of $|M(j\omega)|$ curve is called cutoff rate. Cutoff rate indicates the ability of a system in distinguishing signals from noise. Nyquist criterion, a semi graphical method, originated as engineering application of the well-known "principle of the argument", a concept in complex-variable theory. Then Nyquist Criterion forms basis of all polar plot methods and Bode diagram. Several terms such as: encircled, enclosed, closed path, s-plane, $\Delta(s)$- plane, Γ_s −path, Γ_Δ −path, and number of encirclements, need to be clarified before we delve into principle of argument and Nyquist criterion. In the following, we use a function $\Delta(\cdot)$ of complex variables s, which is single valued i.e., each value of s is associated with one and only one value $\Delta(\cdot)$.

The discussion, to begin with, may appear vague and obscure but if the terms are understood carefully and if the line of thought is followed, we will begin to see in amazement, the stroke of genius with which Nyquist proposed his criterion which became central to all frequency response methods!

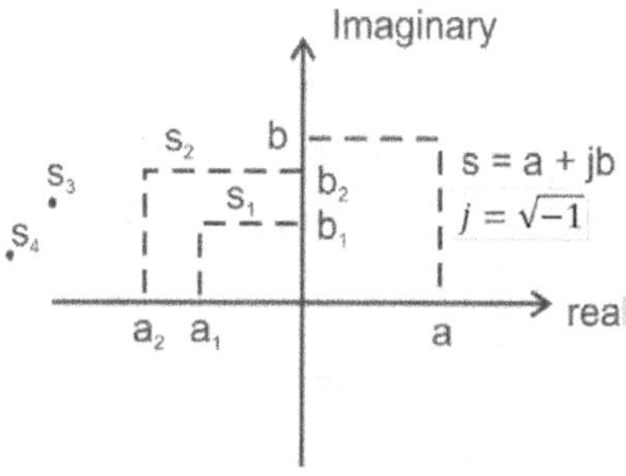

The variable "s", in terms of which characteristic equation is written, is a complex quantity and is plotted on a graph sheet with real and imaginary axis as shown in figure beside. s- plane is the name given to this graph sheet. It may be noted that the variable as is the independent variable and can take values $s = s_1, s_2, \ldots\ldots$ etc. These values of $s, s_1 = a_1 + jb_1, s_2 = a_2 + jb_2$ etc are plotted on s-plane as shown in figure given above.

If a sequence of points $s_1, s_2, \ldots\ldots, s_n, s_1$, so that the starting point and the ending point are the same, are plotted and connected by smooth line, it forms of closed path. This Path is often termed Γ_s −path or Γ_s −curve.

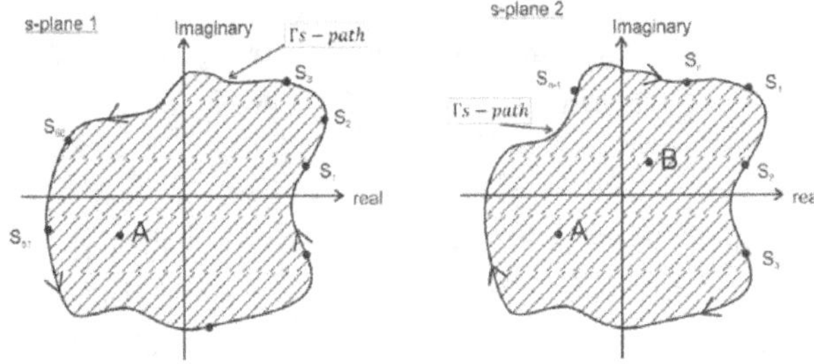

5. Encircled

Two such Γ_s −paths are shown in s- plane 1 and s-plane 2 (i.e., graph sheet 1 and graph sheet 2). Note carefully that when we traverse along the Γ_s −Curve on s-plane 1, and go from s_1 to s_2 to $s_3 \ldots\ldots$ to s_n, we will be going in counter-clockwise direction as viewed from top on to s-plane 1. Similarly, when viewed from top on to s-plane 2, the traversal is in clockwise direction. We say that the point A and all the region inside the curve in s-plane 1 is encircled by the Γ_s-path in counter-clock-wise direction and the points A,B, and all the region inside the curve in s-plane 2 are encircled in clock-wise direction by the Γ_s −path in s-plane 2.

A point or a region in s-plane is said to be <u>encircled</u> by a closed path if it is found inside the path.

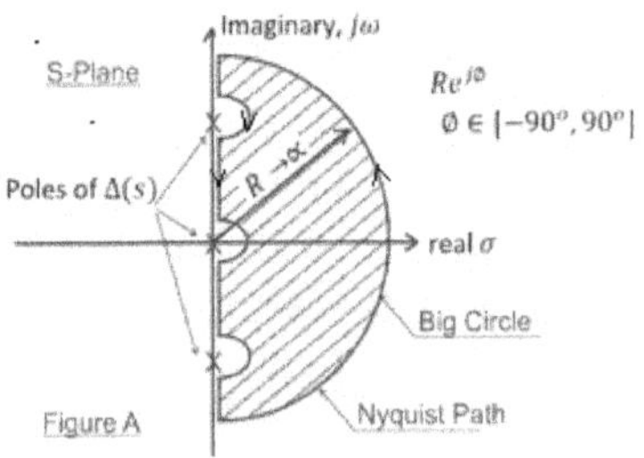

Nyquist chose one particular Γ_s − path which encircles the whole region to the right of imaginary axis. this choice Γ_s −path is called **Nyquist path** as shown in figure. The shaded region, which is entire right half of s- plane, is encircled by <u>Nyquist- Path</u> once in counter-clockwise direction. Before we leave s- plane for a while and go on to next topic, see figure 1,2, to see how to count the member of encirclements.

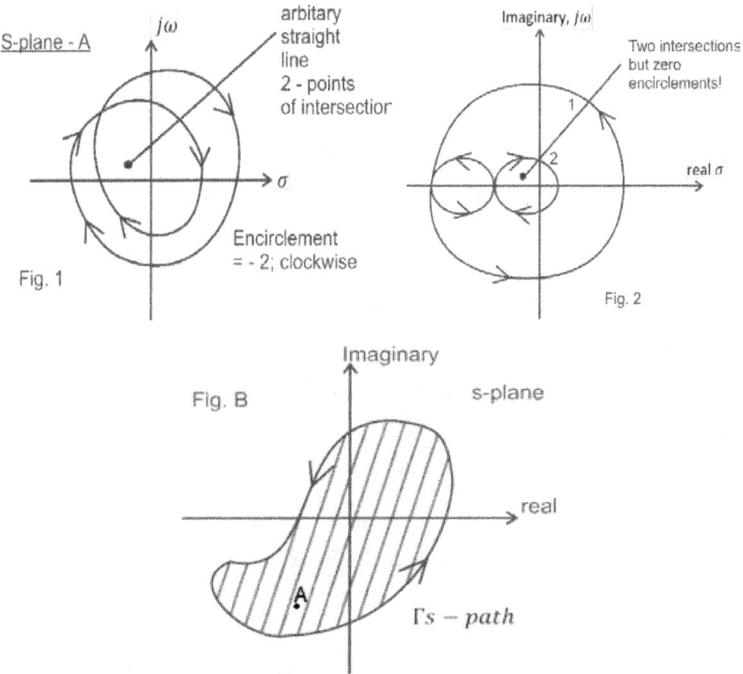

The method of counting encirclement comes in very handy when the map Γ_s −path into Γ_Δ −path as explained later.

6. Enclosed

A point or a region is said to be enclosed by a closed path if it is encircled in the counter-clock-wise direction or the point or the region lies to the left of the path when the path is traversed in the prescribed direction.

When we walk along the Γ_s −path shown besides, the point A and the shaded region are found to lie onto our left hand side. Hence the point A and the hatched region are enclosed by Γ_s − Path.

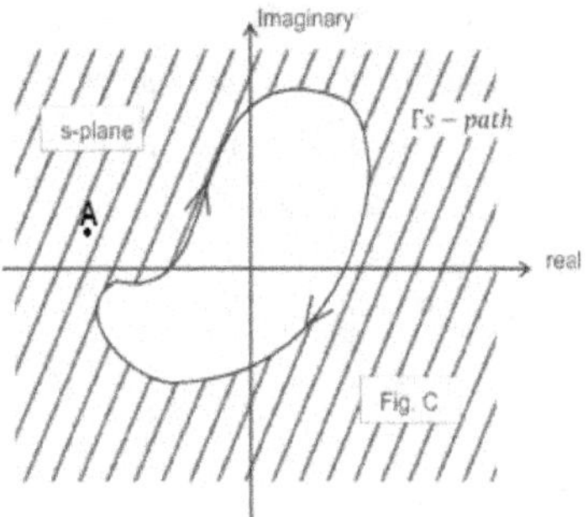

For the Γ_s − path shown in figure C, when it is traversed in the directions shown, the point A, and the region hatched are found to lie on the left hand side. Hence, point A and all the points in the hatched region are "enclosed" by Γ_s − path shown in figure c.

As we will see in subsequent pages, The Nyquist criterion can be specified for minimum-phase-transfer functions by considering whether the Nyquist plot encloses origin (or the $(-1, j0)$ point).

We consider denominator of the transfer function given in equation (A) $\Delta(s) = 1 + G(s)H(s)$ since the roots of $\Delta(s) = 0$ determine the transient response characteristics and the stability of the system. Notice that G(s)H(s) may include terms of the form $e^{-T_d s}$, pure delay to form transfer function of the form.

$$G(s)H(s) = \frac{K(s+z_1)(s+z_2)\ldots\ldots(s+z_m)}{s^P(s+p_1)(s+p_2)\ldots\ldots(s+p_n)} e^{-Tds} \ .$$

7. Mapping

Let us consider a Γ_s −path in s-plane and choose points $s_1, s_2 \ldots\ldots$ along the Γ_s −path as shown in figure below.

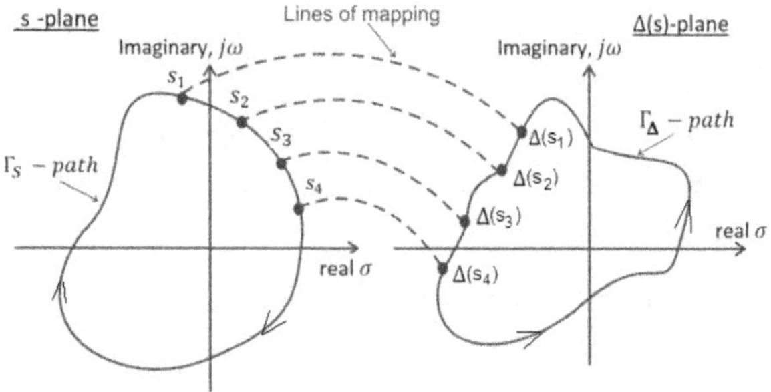

Corresponding to each points on the Γ_s −Path, compute the $\Delta(s)$ value, and plot it on another graph sheet. This second graph sheet is termed $\Delta(s)$- plane. That is, for $s = s_1$, compute $\Delta(s_1)$, for $s = s_2$, compute $\Delta(s_2)$ and so on. Since s_1, s_2 etc. are complex quantities $\Delta(s_1), \Delta(s_2) \ldots\ldots$ etc are also complex quantities. When we plot $\Delta(s_1), \Delta(s_2) \ldots\ldots$etc on $\Delta(s)$ −plane and join by smooth curve, we obtain Γ_Δ −path.

The lines of mapping are shown to indicate that for $s_1, s_2, s_3 \ldots\ldots \Delta(s_1), \Delta(s_2)\Delta(s_3)$ are computed and also to specify that for each s, there is only one $\Delta(s)$ or $\Delta(s)$ is a single valued function. Other than that, there is no necessity for showing lines of mapping. Notice that, in this case, Γ_s − path is shown to traverse in clock-wise direction whereas Γ_Δ −path is shown to traverse in counter-clock-wise direction. Whether the Γ_Δ −path is traversing in clock-wise or counter-clock-wise direction depends on the function $\Delta(s)$ and the Γ_s − path chosen.

What we have seen so far is that we consider a Γ_s −path and through mapping process, plot Γ_Δ − path. When we apply the principle of argument in the form of Nyquist criterion, we choose one particular choice of Γ_s −path, termed as Nyquist path, as shown in figure A. We choose $\Delta(s)$ at the denominator of transfer function in equation (A). Then, obtain the Γ_Δ −path on $\Delta(s)$plane.

Note clearly that when the Γ_s −path is chosen to be the **Nyquist path**[19] shown in Figure A, the corresponding Γ_Δ −path obtained by the process of mapping is called the **Nyquist plot**.

Now we consider the principle of argument and then come back to the problem of stability.

1.23. Principle of Argument

Let $\Delta(s)$ be a single-valued function that has finite number of poles in the s-Plane. Suppose that an arbitrary closed path Γ_s −path is chosen in the s-plane so that the path does not go through any one of the poles or zeros of $\Delta(s)$; the corresponding Γ_Δ −path mapped in the $\Delta(s)$ −plane will encircle the origin of $\Delta(s)$ −plane as many times as the difference between the number of zeros and poles of $\Delta(s)$ that are encircled by the s- plane path Γ_s −path.

Carefully see the process implied in the principle of argument:

1) Choose a $\Delta(s)$ function. We choose it as the denominator of the transfer function given in Equation (A). We know the poles of $\Delta(s)$ function.

2) Choose a Γ_s −path . Draw it on s-plane. We choose it as Nyquist path to cover the entire right hand side of imaginary axis as shown in Figure A. Plot the poles of on s-plane.

3) Count how many of the poles are encircled by Γ_s −path. Let this number be P.

---

[19] Clearly understand the terms "Nyquist path" and "Nyquist plot" before going on any further.

4) Through the process of mapping, plot the Γ_Δ —path on $\Delta(s)$-plane.

5) Count how many times the origin of $\Delta(s)$-plane is encircled by the Γ_Δ —path. Let this number be N.

6) As per principle of argument, N=Z - P where Z is the number of zeros of $\Delta(s)$ encircled by Γ_s —path .

7) Z can be easily obtained. This is the number of zeros of $\Delta(s)$ enclosed by Γ_s —path.

8) The way $\Delta(s)$ is defined in Equation (A), zeros of $\Delta(s)$ are roots of characteristic equation. Thus, when we compute Z, we are computing the number of roots of characteristic equation that are encircled by the Γ_s —path.

9) With the choice of Γ_s —path shown in Figure A, we want Z to be zero.

10) If we substitute Z=0, Nyquist criterion implies N=-P.

A bit more explanation of the ten points mentioned is offered in the following. Follow through it patiently. For motivating Nyquist criterion, let $\Delta(s) = 1 + G(S)H(S) = 1 + \frac{N_1(s)}{D_1(s)}$ so $\Delta(s) = \frac{N_1(s)+D_1(s)}{D_1(s)}$. Notice two things: (i) $N_1(s) + D_1(s) = 0$, that is zeroes of $\Delta(s)$, are roots of characteristic equation. (ii) $D_1(s) = 0$ gives poles of $\Delta(s)$ as well as G(s)H(s).

We do NOT know the zeros of $\Delta(s)$ but we know poles of $\Delta(s)$ since they are the same as poles of G(s)H(s).

If we considered the Γ_s —Path as the Nyquist path as shown in figure A , then, we can easily find the number of poles of $\Delta(s)$ encircled by Nyquist path, because these are the poles of G(s)H(s) lying in right half of s-plane. Let this be P. We plot $\Delta(s)$ and obtain number of encirclement of origin by Γ_Δ —Path (Γ_Δ —path for Nyquist part is called Nyquist plot). Let N be the number of encirclement of origin of $\Delta(s)$-plane by Γ_Δ —path . Then, according to principal of argument, N = Z-P. where Z is the number of zeros of $\Delta(s)$ encircled by Γs —path or number of zeros of $\Delta(s)$ lying in right half of s- plane. For stability Z should be zero. $\therefore \boxed{N = -P}$

Procedure for Applying Nyquist Criterion

The procedure for applying Nyquist Criterion is listed once again here for reinforcing the idea.

1. Choose $\Delta(s) = 1 + G(s)H(s) = 1 + \frac{N_1(s)}{D_1(s)} = \frac{N_1(s)+D_1(s)}{D_1(s)}$

Zeros of $\Delta(s)$: $N_1(s) + D_1(s) = 0$: same as roots of char. Eqn. poles of $\Delta(s)$: same as poles of G(s)H(s)

We need to find how many roots of characteristic equation lie and RHS of s- plane: that is how many zeros of $\Delta(s)$ lie in right half of s- plane. For stability there should not be any zeros of $\Delta(s)$ in right half of s- plane.

2. Draw the Nyquist Path on s- plane. On the same s- plane, locate the poles of G(s)H(s). Count the poles of $\Delta(s)$ (or poles of G(s)H(s)) that are encircled by Nyquist path. let this number be P.

3. Now, draw Γ_Δ −path (which is Nyquist plot) for several points on Γ_s −Path. Count the number of encirclement of origin using the concept specified in figures 1 and 2. Let this number be N.

4. According to the principle of argument: N = Z - P. For stability N = - P.

Nyquist Criterion Specified

For a closed loop system to be stable, the $\Delta(s)$ plot (i.e., nyquist plot) must encircle the origin of $\Delta(s)$plane as many times as the number of poles of $\Delta(s)$ that are in the right half of s-plane, and the encirclement, if any, must be made in the clockwise direction (if γs −path is traversed in the clockwise direction).

The origin of $\Delta(s)$plane is called the critical point because we are interested in finding the encirclement of the Γs −plot around the origin.

The procedure seems fine except for step (iii) which can be further simplified.

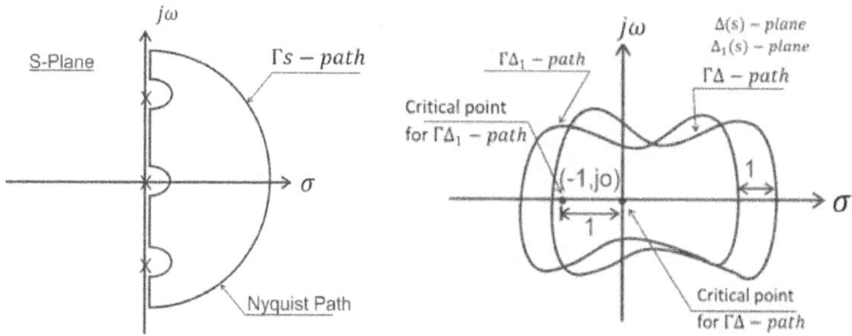

Consider the Γ_Δ −path drawn on $\Delta(s)$ −plane. Define $\Delta_1(s) = \Delta(s) - 1$. And draw the Γ_{Δ_1} −path. Subtracting 1 from $\Delta(s)$ has the effect of moving every point on Γ_Δ −path 1 unit to the left. Thus, Γ_{Δ_1} −path is obtained by moving the Γ_Δ −path 1 units to the left. In this process, the critical point is also shifted by 1 unit to the left. Finally (and once and for all) we define Γ_{Δ_1} −path to be Nyquist plot. Note that $\Delta(s) = 1 + G(s)H(s)$ and $\Delta_1(s) = G(s)H(s)$. With this, we specify the Nyquist Criterion as:

Simplified Nyquist Criterion

For a closed loop system to be stable, the Nyquist plot for the open loop transfer function must encircle the $(-1, j0)$ point as many times as the number of poles of $G(s)H(s)$ that are in the right-half-s-plane, and the encirclements, if any, must be made in the clockwise direction (if Γ_s — path traversed in the counter clockwise direction).

See the stroke of genius of H. Nyquist: by a graphical approach used on open loop transfer function, stability of the closed loop system is analyzed by employing the very well-known principle of argument of complex variable theory. It took several years and the genius of Nyquist for this wonderful result to be presented to the world.

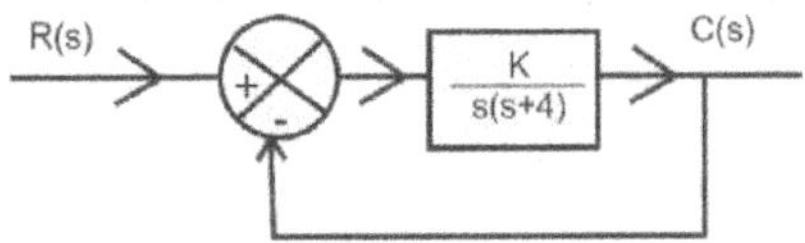

Example 1: Nyquist Criterion

Using the Nyquist criterion, comment on the stability of closed loop system shown in figure.

[This example is perhaps a very simple one and using Nyquist Criterion for this problem may be akin to throwing a hydrogen bomb to get rid of a sparrow! But we will go through the steps of the Nyquist Criterion just for the sake of explaining the principle and showing an example.]

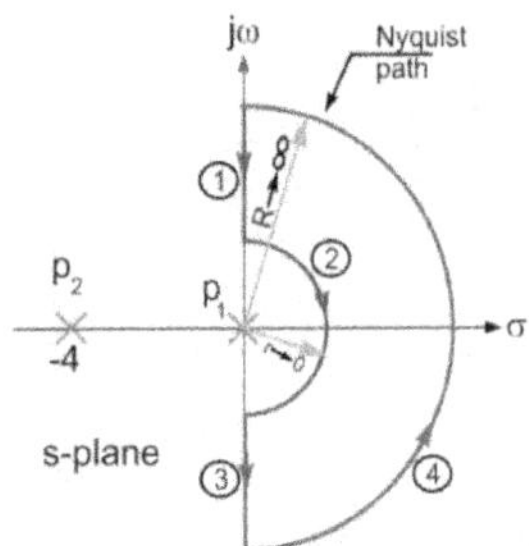

<u>Solution</u>:- Consider the Nyquist path shown. Notice that the small semicircle at origin is considered to ensure that Nyquist path does not pass through pole at origin. Notice that there are four portions of the Nyquist path:

(1) Portion on imaginary axis from $j\omega \to \infty$ to $j\omega \to 0$.

(2) Small semi-circle $r \to 0, \theta$ varying from $90°$ to $-90°$.

(3) Portion on imaginary axis from $j\omega \to 0$ to $j\omega \to -\infty$.

(4) Big semi-circle $R \to \infty, \theta$ varying from $-90°$ to $+90°$.

With the OLTF $G(s)H(s) = \frac{K}{s(s+4)}$ and nyquist path given, we see that none of the poles of $G(s)H(s)$ encircled by Nyquist path. $\therefore$ P = 0.

Now, let us draw Nyquist plot for G(s)H(s). For convenience and ease of reference, let us consider four parts of the Nyquist path as shown by (1), (2), (3) and (4). We map the s-plane path into Δ_1(s)-plane path and obtain Nyquist plot corresponding to each part.

Part 1 of s-plane path:- $s = j\omega$ and ω goes from ∞ to zero on s-plane.

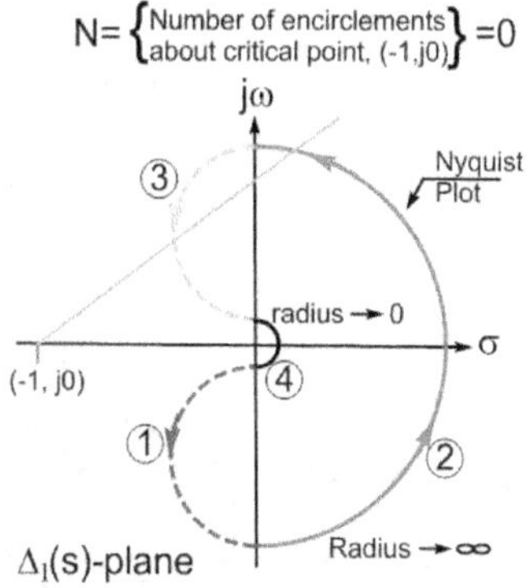

$$G(s) = \frac{K}{j\omega\,(j\omega +4)} \Rightarrow |G(s)| = \frac{K}{\omega\sqrt{\omega^2+16}} \text{ and}$$

$$\angle G(s) = -90 - \tan^{-1}\left(\frac{\omega}{4}\right)$$

$$\omega \to \infty \Rightarrow \underbrace{|G(s)| \to 0}_{begins}, \angle G(s) \to -180^o$$

$$\omega \to 0 \Rightarrow \underbrace{|G(s)| \to \infty}_{ends}, \angle G(s) \to -90^o$$

Carefully notice where does portion begin and end on Δ_1(s)-plane. It begins very close to origin($|G(s)| \to 0$) on the negative real axis where angle is -180° ($\angle G(s) \to -180^o$). And it ends far, far away from origin($|G(s)| \to \infty$) and near negative imaginary axis at angle -90°($\angle G(s) \to -90^o$).

Portion 2 of s-plane path:- $s = re^{j\emptyset}; r \to 0, \emptyset \to +90^o$ to -90° on s-plane.

$$\therefore G(s) = \frac{K}{re^{j\emptyset}(\underbrace{re^{j\emptyset}}_{\substack{\text{very small as} \\ \text{compared to 4} \\ \text{since r}\to 0}} + 4)} \approx \frac{K}{4r}e^{-j\emptyset}$$

$\therefore |G(s)| \to \infty$ and $\underbrace{\emptyset \to -90^\circ}_{begins}$ to $\underbrace{+90^\circ}_{ends}$ on $\Delta_1(s)$-plane. That is the Nyquist path starts far, far

away from origin of on $\Delta_1(s)$-plane ($|G(s)| \to \infty$) and moves from -90° to $+90^\circ$.

Portion 3 of s-plane path:- Is mirror image of (1) about real axis.

Portion 4 of s-plane path:- $s = Re^{j\emptyset}, R \to \infty, \emptyset \to -90^\circ$ to 90° on s-plane.

$$G(s) = \frac{K}{Re^{j\emptyset}(Re^{j\emptyset} \underbrace{+4}_{\substack{\text{may be}\\\text{ignored}\\\text{since } R\to\infty}})} = \frac{K}{R^2}e^{-j(2\emptyset)}$$

$$\therefore |G(s)| \to 0, \emptyset \to 180^\circ \text{ to } -180^\circ$$

Since N = - P, the system is stable for all values of K.

Nyquist Criterion for Minimum Phase Systems[20]

The Nyquist Criterion becomes much simpler for system with minimum phase transfer function. Majority of the loop transfer function encountered in real-world are of minimum phase type. Nyquist Criterion for systems with minimum phase transfer function is considerably simple and is stated as shown below:

For closed loop system with loop transfer function L(s) that is of minimum phase type, the system is closed loop stable if the L(s) plot for s = jω and ω varying from ∞ to zero does not enclose the (−1, j0) point.

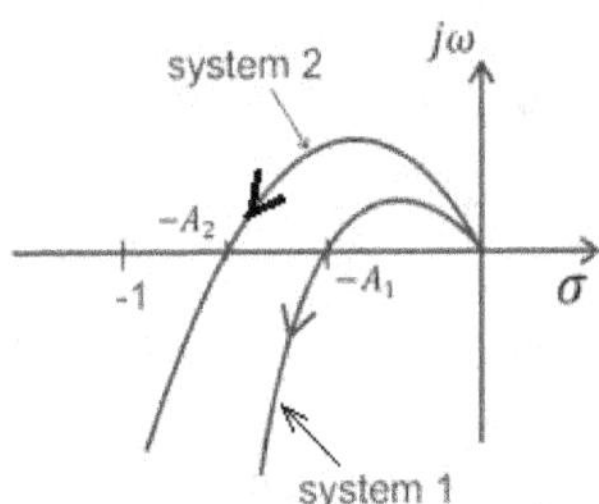

As seen from Example 1, Nyquist Criterion in general is Semi-graphical in the sense that we did not plot the Nyquist plot to scale. However, the Nyquist Criterion for systems with minimum phase transfer function is amenable for mathematical analysis as well as polar plot analysis and Bode diagram as will be discussed subsequently. Further, the Nyquist Criterion for systems with minimum phase transfer function can be used to develop the concept of gain margin and phase margin, Bode diagram also stems from this criterion.

---

[20] What is a minimum phase system? Find out and let me know.

To motivate this discussion, let us say we have two systems and we plotted the plot for both these systems as shown in the Figure. Let these plots intersect the negative real axis in -A_1 and -A_2. Since A2 > A1, we say that system 2 is less stable as compared to system 1. We will generalize this into the concept of ***gain margin*** in subsequent discussion.

Note that both the systems are stable since both the plots do not enclose (-1,j0) point since the (-1,j0) point does not lie to the left when the polar plot is traversed in the direction specified. This, then forms a condition for stability. If the point of intersection on the negative real axis has a magnitude of less than 1 the system is stable. If the magnitude is less than 1, the system is unstable. If it is equal to 1, the system is marginally stable.

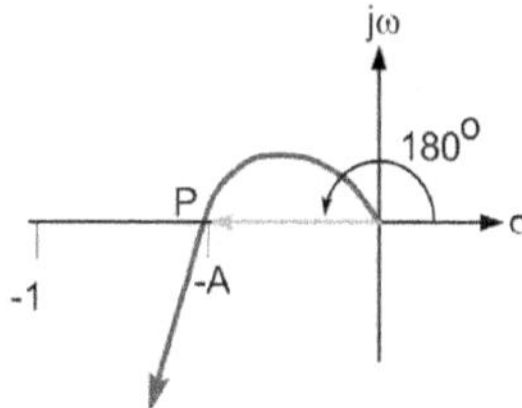

Example 2: Analytical Method

The open loop transfer function of a system is:

$$G(s)H(s) = \frac{100}{s(s + 2)(s + 10)}$$

Determine if the closed loop system is stable.

Solution

To apply the simplified Nyquist criterion for minimum phase systems, we first observe the Nyquist plot for part 1 of the Nyquist path. Fortunately, we do not have to plot it on graph sheet to scale. We will infer the stability mathematically. In the polar plot shown, if $A < 1$, the critical point, $(-1, j0)$ Is not encircled by the Nyquist plot and the system is stable. We will now try to compute the value of A.

At point P on the polar plot of the system,

$$|G(s)H(s)| = A \quad \text{(I)}$$
$$\angle G(s)H(s) = 180° \quad \text{(II)}$$

The second condition on the angle can be used to find the frequency at point P.

$$G(j\omega)H(j\omega) = \frac{100}{j\omega(j\omega + 2)(j\omega + \omega)} = \frac{100}{-12\omega^2 + (20\omega - \omega^3)j}$$

$$\therefore \angle G(j\omega)H(j\omega) = -tan^{-1}\left(\frac{20\omega-\omega^3}{\omega^2}\right) = 180° \Rightarrow \omega = \sqrt{20}\ \text{rad/sec}$$

$$\therefore G(j\omega)H(j\omega) = \frac{100}{\omega\sqrt{\omega^2+4}\sqrt{\omega^2+100}} = 0.41 = A$$

Since $A < 1$, the system is stable.

The conditions (I) & (II) may also be specified as:

Real part of $G(j\omega)H(j\omega) = -A$, and

Imaginary part of $G(j\omega)H(j\omega) = 0$ to get the same result.

Same result can be obtained

from Routh criterion also.

$$s^3 + 12s^2 + 20s + 100 = 0$$

$$
\begin{array}{llll}
s^3 & : & 1 & 20 \\
s^2 & : & 12 & 100 \\
s^1 & : & \dfrac{140}{12} & \\
s^0 & : & 100 &
\end{array}
$$

No sign change. So stable

Example 3: System with pure time delay

The block diagram of a closed loop system is shown in figure below:

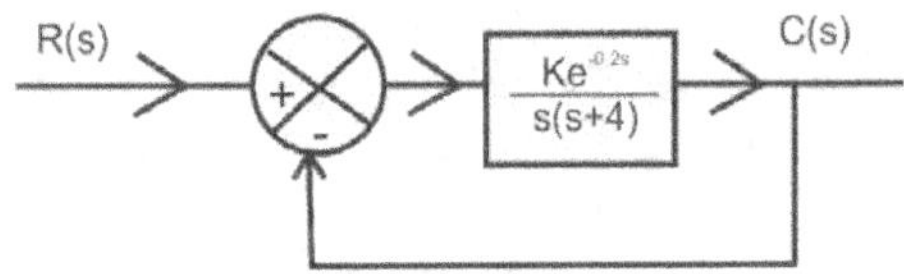

Find the maximum value of the system gain K which will keep the closed loop system.

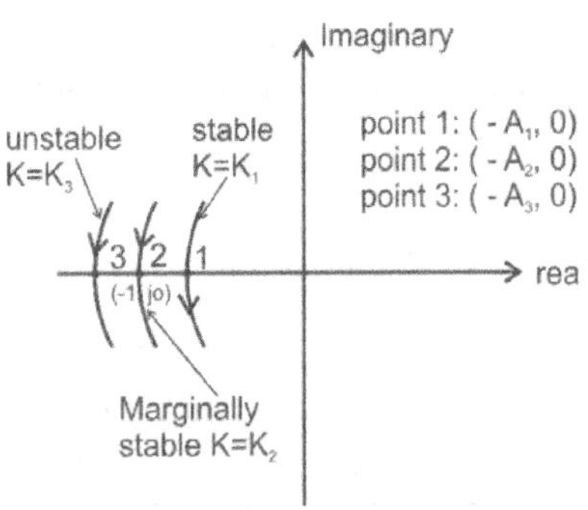

Solution:- The forward path transfer function has a term $e^{-0.2s}$. This term corresponds to pure time delay. Neither Routh's Criterion nor root-locus method can solve touch problem without resorting to approximations. Nyquist Criterion can effectively give a solution without resorting to any approximations. The method is a bit computation intensive, though.

$$\text{Let } G(s) = \frac{Ke^{-0.1s}}{s(s+4)}.$$

We note that, if K is changed, let us say by a factor of x, then every point, including the point of intersection on the negative real axis, get multiplied by the same factor.

Figure besides shows sketches of Nyquist plot of $G(s)$ near the point of intersection with negative real axis for $K = K_1, K_2$ and K_3 respectively. When $K = K_1$, the system is stable since the critical point $(-1, j0)$ is NOT enclosed. When value of K is increased by a multiple of, say x, to get $K = K_1 x = K_2$, the system is on the borderline of stability and instability (called marginally stable) and the curve passes through the critical point. When the value of K is further increased to $K_3 = K_1 y$ where $y > x$, the system becomes unstable since the critical point is enclosed. When $K = K_2, A_2 = A_1 x = 1$.

[Since the coordinates of every point on curve for $K = K_1$ are multiplied by a factor x, coordinates of (-1,j0) are also multiplied by a factor x. $\therefore A_2 = A_1 x = 1$. Thus, $x = \frac{1}{A_1}$.]

The procedure involved settings $K_1 = 1$, find the x co-ordinate of point of intersection, $(-A_1, jo)$. And then, find the factor $x = \frac{1}{A_1}$, and find maximum value of $K = K_1 x = \frac{K_1}{A_1}$.

The critical step is to find point 1 on the plot just shown for K=K₁, the intersection of Nyquist plot with negative real axis. This is done by setting:

- Imaginary parts of $G(j\omega) = 0$ and find ω_p
- Find Real part of $G(j\omega)$ at $\omega = \omega_g$.
- This Real part has magnitude of A_1

Set $K = K_1 = 1$ so that $G(s) = \dfrac{e^{-0.2s}}{s(s+4)} \Rightarrow G(j\omega) = \dfrac{e^{-j(0.2\omega)}}{j\omega(j\omega+4)}$

$$\text{Or } G(j\omega) = \frac{[\cos(0.2\omega) - j\sin(0.2\omega)]}{-\omega^2 + j(4\omega)}$$

$$\Rightarrow G(j\omega) = \frac{[\cos(0.2\omega) - j\sin(0.2\omega)][-\omega^2 - j(4\omega)]}{[-\omega^2 + j(4\omega)][-\omega^2 - j(4\omega)]} = -U + jV$$

Where

$$U = \frac{\omega^2\cos(0.2\omega) + 4\omega\sin(0.2\omega)}{\omega^4 + 16\omega^2} \text{ , and}$$

$$V = \frac{\omega^2 \sin(0.2\omega) - 4\omega \cos(0.2\omega)}{\omega^4 + 16\omega^2}$$

$$V = 0 \Rightarrow \omega^2 \sin(0.2\omega) - 4\omega \cos(0.2\omega) = 0$$

$$\Rightarrow f(\omega) = \omega \sin(0.2\omega) - 4\cos(0.2\omega) = 0 \because \omega \neq 0$$

Solve $f(\omega) = 0$ by Bisection Method

ω	0	5	2.5	3.75	4.375	4.062	3.906	3.984	3.945
$f(\omega)$	-4	2.05	-2.31	-0.37	0.794	0.198	-0.0895	0.053	0.018

$\therefore \omega = 3.945 \text{ rad/sec} = \omega_p$

$$\therefore A_1 = U = \frac{\omega_p^2 \cos(0.2\omega_p) + 4\omega_p \sin(0.2\omega_p)}{\omega_p^4 + 16\omega_p^2} = 0.045 \Rightarrow x = \frac{1}{A_1}$$

$\therefore$ Maximum value of $K = \dfrac{1}{A_1} = 22.22$

Students are advised to follow each step and ensure that they can solve this problem on their own.

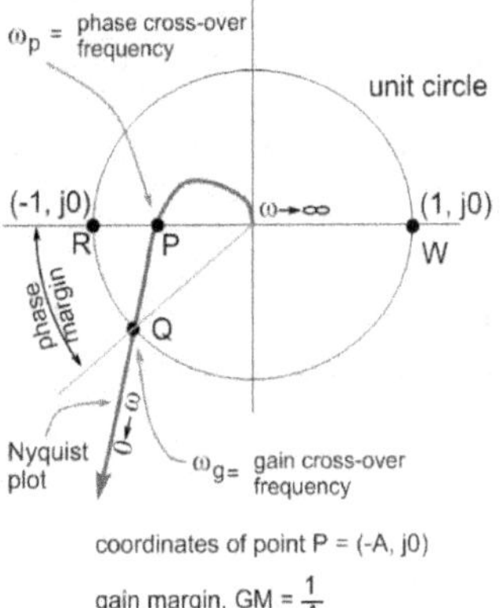

Corresponding to each point on Nyquist plot, there is a frequency ω, magnitude $|G1(j\omega)|$.

The Nyquist Criterion can also be applied graphically and this viewpoint give the notion of gain margin (GM) and phase margin (PM) as shown in the figure beside.

The point where Nyquist plot intersects negative real axis is called phase crossover point and the frequency corresponding to this point is called phase-cross-over frequency, ω_p. Similarly the point where the Nyquist plot intersects the unit circle is called gain crossover point and the corresponding frequency is called gain crossover frequency, ω_g.

Gain Margin is the amount of Gain in decibels (dB) that can be added to the loop before the closed loop system becomes unstable. If the Nyquist plot is drawn for $K = K_1$ and $GH(s) = \frac{K_1(s+z_1)........(s+K_m)}{(s+p_1)......(s+p_1)}$ and phase crossover frequency is ω_p, then gain margin is obtained as:

$$GM = 20\ log_{10}\left[\frac{1}{|GH(j\omega)|}\right]_{\omega=\omega_p} = -20log_{10}|GH(j\omega_p)|\,.$$

Once GM is computed, maximum value of Gain K_{max} may be computed as:

$$20log_{10}\left[\frac{K_{max}}{K_1}\right] = GM \Rightarrow K_{max} = K_1 * 10^{(GM/20)}\,.$$

Phase margin is defined as the angle is degrees through which the Nyquist plot must be rotated about the origin so that the gain crossover passes through the $(-1, j0)$ point.

If ω_g is the gain crossover frequency then phase margin (pm) is

$$PM = GH(j\omega_g) - 180^\circ\,.$$

To summarize, given GH(s):

gain crossover frequency ω_g is computed from: $|GH(j\omega_g)| = 1$

phase crossover frequency ω_p is computed from: $\angle GH(j\omega_p) = 180^\circ$

gain margin is computed as: $GM = -20\ log_{10}|GH(j\omega_p)|$ dB

Phase margin is computed as: $PM = \angle GH(j\omega_g) = -180^\circ$

Nyquist criterion, when used as a graphical method, is very useful tool. But one limitation is that frequency information is not directly visible In the Nyquist plot. Other limitations include the difficulties in drawing the Nyquist plot to scale and the difficulties in designing compensators. All these difficulties are addressed in Bode diagrams. Bode diagram includes two plots: (i) magnitude (in dB). Vs. frequency and (ii) Phase in degrees. Vs. frequency. In both the plots, the horizontal axis is frequency in logarithmic scale and same ω values are used for both the plots.

Bode plots are preferred for the following reasons:

a) Bode diagram can be quickly drawn by approximating the magnitude and phase with straight line segments

b) GM, PM, gain crossover, and phase crossover are very easily determined from Bode plot than Nyquist plot

c) Effects of adding controllers and their parameters are easily visualized on Bode plot then on Nyquist plot.

d) When experimental frequency response data is available for magnitudes of input and output, it is possible to arrive at an estimate of transfer function using Bode plot.

We will see one example of drawing exact Bode plot and 1 example of drawing approximate Bode plot.

Example 4: Gain Margin and Phase Margin

Comment on the stability of the system with following block diagram:

Solution:- $GH(s) = \dfrac{2500}{s(s+5)(s+50)} \Rightarrow GH(j\omega) = \dfrac{2500}{j\omega\,(j\omega+5)(j\omega+50)}$

$\therefore |GH(j\omega)| = \dfrac{2500}{\omega\sqrt{\omega^2+25}\sqrt{\omega^2+2500}}$ and so $|GH(j\omega)|_{dB} = 20\,log_{10}|GH(j\omega)| = M$

$\angle GH(j\omega) = -90 - tan^{-1}\left(\dfrac{\omega}{5}\right) - tan^{-1}\left(\dfrac{\omega}{50}\right) = \varnothing$

Table below is constructed showing M_{dB} and $\varnothing$ Degrees.

ω	0.01	0.05	0.1	0.5	1	5	10	30
M_{dB}	60	46	40	25.97	19.83	2.96	-7.16	-26.56
$\varnothing$ degree	-90.13	-90.6	-91.28	-96.28	-102.46	-140.71	-164.74	-201.56

As shown on graph, system is stable and remains stable till a value of $K_{max} \cong 15775$

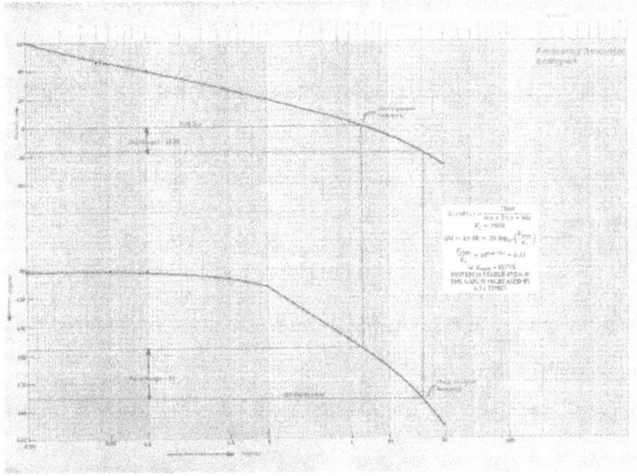

- The phase and gain margins of a control system are a measure of the closeness of polar plot to the $-1 + j0$ point.

- GM or PM alone may not give a sufficient indication of the relative stability

- For a minimum phase system, both GM & PM must be positive for system to be stable.

- Proper GM & PM ensure stability in the face of variations in the system components up to an extent.

- The two values GM & PM "bound" the behavior of the closed loop system near resonant frequency

- Generally GM is chosen greater than 6dB and PM is chosen to be between 30 degree and 60 degree.

- In most situations slop of -20 dB/ decade is desirable at the gain crossover frequency

- If it is - 40 dB/ decade at ω_g, generally very small GM will result in systems that are very close to stability and instability.

- - 60 dB/decade will most probably make the system is unstable

- For non minimum phase systems, it may be a good idea to use Nyquist Criterion and Nyquist diagram to determine the stability rather than bode diagram approach since it takes careful study for the correct interpretations of stability margins.

Approximate Magnitude Plot

Magnitude and phase plots of bode diagram may be drawn using *asymptotic approximation*. Though there is an error involved in approximation, it may be desirable to have a quick estimate of GM & PM while performing design iterations. In such situations, asymptotic approximation finds its use. When once approximate calculations provide a "solution", it may be fine-tuned using computer programs such as MATLAB.

Consider a system with open loop transfer function.

$$G(s)H(s) = \frac{K(s + z_1)(s + z_2) \dots \dots (s + z_m)}{s(s + p_1)(s + p_2) \dots \dots (s + p_n)}$$

For now, we consider only real poles and zeros and write terms in the form:

$(s + z_1) = z_1 \left(1 + \frac{s}{z_1}\right), (s + p_1) = p_1 \left(1 + \frac{s}{p_1}\right)$ etc to obtain:

$$H(s)G(s) = \frac{Kz_1 \left(1 + \frac{s}{z_1}\right) z_2 \left(1 + \frac{s}{z_2}\right) \dots \dots z_m \left(1 + \frac{s}{z_m}\right)}{sp_1 \left(1 + \frac{s}{p_1}\right) p_2 \left(1 + \frac{s}{p_2}\right) \dots \dots p_n \left(1 + \frac{s}{p_n}\right)} = \frac{K_1 (1 + T_1 s) \dots \dots (1 + T_m s)}{s(1 + \tau_1 s) \dots \dots (1 + \tau_n s)}$$

where $K_1 = \frac{Kz_1 z_2 \dots \dots z_m}{p_1 p_2 \dots \dots p_n}, T_1 = \frac{1}{z_1}, \dots \dots, T_m = \frac{1}{z_m}, \tau_1 = \frac{1}{p_1}, \dots \dots, \tau_n = \frac{1}{p_n}$

The open loop TF now has terms of three forms: (i) constant (ii) lone -s- term, and (iii) term of the form $(1 + Ts)$. For the constant term and lone-s-term, approximation is not necessary. Approximation for the first order terms, $(1 + Ts)$ is shown here.

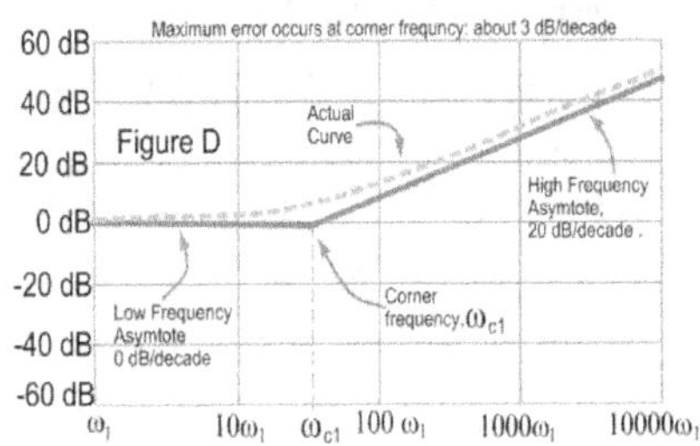

(1 + Ts) term in numerator: Asymptotic Approximation of magnitude plot.

Magnitude plot: $(1 + Ts)$ in Numerator

Consider the term $F_1(s) = (1 + T_1 s)$ appearing in numerator.

$$F_1(j\omega) = (1 + jT_1\omega) \text{ and } |F_1(j\omega)| = \sqrt{1 + T_1^2\omega^2}.$$

Noting this, we can see that when ω is very small $\omega \to 0$, $|F_1(j\omega)| \approx 1$, and when ω is very large $\omega \to \infty$, $|F_1(j\omega)| \approx T_1\omega$. This gives us the idea of **low frequency asymptote** and **high frequency asymptote**. When written in dB:

$$|F_1(j\omega)|_{dB} = 20log_{10}|F_1(j\omega)|_{dB} = 20log_{10}1 = 0 \text{ along low frequency asymptote}$$
$$= 20log_{10}|G(j\omega)| = 20log_{10}(T_1\omega) \text{ along high frequency asymptote}$$

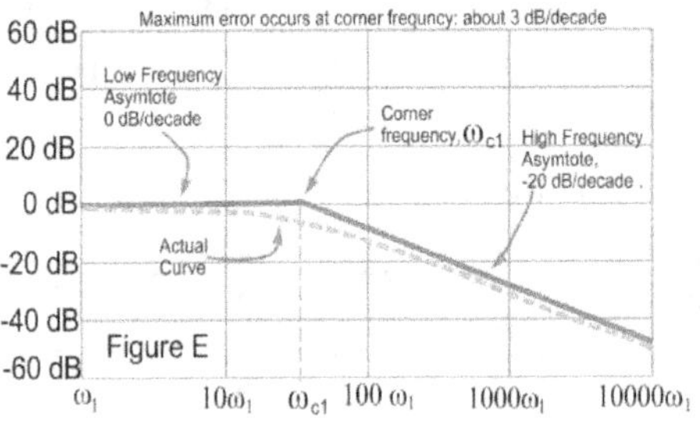

(1 + Ts) term in denominator: Asymptotic Approximation of magnitude plot.

The approximate plot may be shown as in Figure D. The magnitude plot for $(1 + T_1 s)$ in numerator is approximated by two straight line segments: zero dB until ω_{c1} and 20 dB/decade after ω_{c1}. ω_{c1} is termed the **corner frequency**: $\omega_{c1} = \frac{1}{T_1}$. The maximum error of approximation occurs at corner frequency and is about 3dB.

Magnitude Plot: $(1 + \tau s)$ *In Denominator*

Thinking along the same lines,

$$F_1(j\omega) = \frac{1}{1 + j\tau_1\omega} \Rightarrow |F_1(j\omega)| = \frac{1}{\sqrt{1 + \tau_1^2\omega^2}}$$

And $|F_1(j\omega)|_{dB} = -20\log_{10}\sqrt{1 + \tau^2\omega^2}$ gives

The approximate plot is shown in Figure E. Since the term is in denominator, the slope happens to be – 20 dB/decade

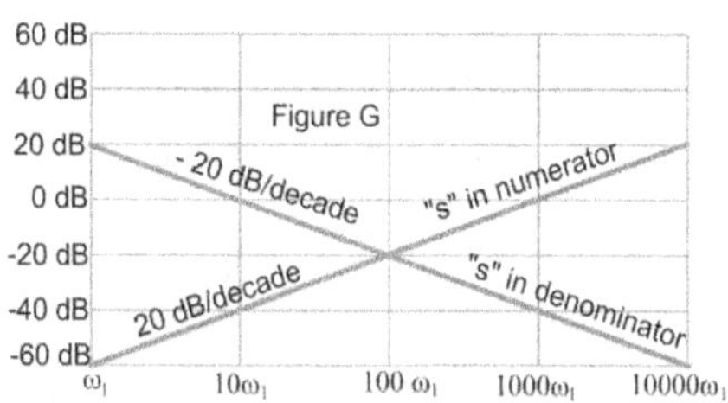

No approximation is involved in plotting a lone s-term. In an actua sitaution, the dB values plotted will depend on the choice of frequency range.

Lone s-Term

Returning back to the transfer function

$$H(j\omega)G(j\omega) = \frac{K_1(1+T_1s)(1+T_2s)......(1+T_m s)}{s^q(1+\tau_1s)(1+\tau_2s)......(1+\tau_n s)},$$

We see that if q, the power of loan-s-term is positive, s^q is a term in the denominator; if it is negative, s^{-q} is a term in the numerator. For now, say $q = 1$: a single pole at origin. This term shows up as a line segment for magnitude plot with slope of - 20 dB/ decade. Similarly, if s appears in numerator, a single zero at origin, the line segment has a slope of + 20dB/ decade.

Constant Term

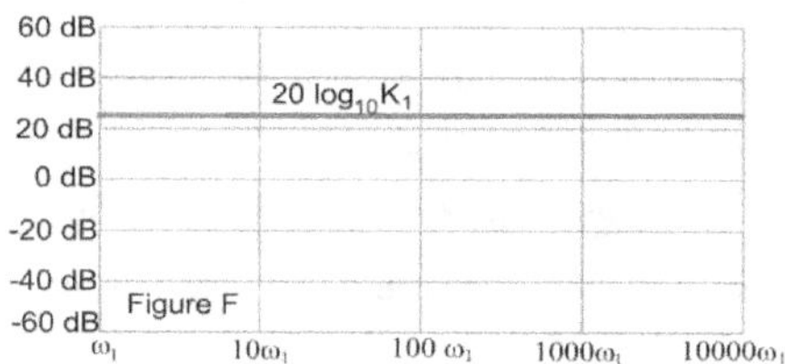

No approximation is involved in plotting a constant term.

The K_1 −term has constant contribution at a height of 20 $\log_{10}K_1$ the line in Figure F is above 0 dB line if K1 > 1 or else it will be below 0 dB line.

With this said, we write

$$|G(j\omega)H(j\omega)| = \frac{K_1(1+jT_1\omega)(1+jT_2\omega)(1+jT_3\omega)\dots\dots(1+T_m\omega)}{(j\omega)(1+\tau_1\omega)(1+\tau_2\omega)(1+\tau_3\omega)\dots\dots(1+\tau_n\omega)}$$

$$= \frac{K_1\sqrt{1+T_1^2\omega^2}\sqrt{1+T_2^2\omega^2}\dots\dots\sqrt{1+T_m^2\omega^2}}{\omega\sqrt{1+\tau_1^2\omega^2}\sqrt{1+\tau_2^2\omega^2}\dots\dots\sqrt{1+\tau_n^2\omega^2}}.$$

And $|G(j\omega)H(j\omega)|_{dB} = 20log_{10}|G(j\omega)H(j\omega)|$

$$= \overbrace{20log_{10}K_1}^{\text{See Figure F}} + \overbrace{20log_{10}\sqrt{1+\tau_1^2\omega^2} + \dots\dots\dots + 20log_{10}\sqrt{1+\tau_m^2\omega^2}}^{\text{First order.See Figure D}} \underbrace{- 20log_{10}\omega}_{\text{See Figure G}}$$

$$- \underbrace{20log_{10}\sqrt{1+\tau_1^2\omega^2} + \dots\dots\dots}_{\text{First order.See Figue E}} \underbrace{20log_{10}\sqrt{1+\tau_n^2\omega^2}}_{\text{First order.See Figue E}}$$

Since each term is approximated by straight line segments, the whole plot is obtained as a set of approximate line segments. This process may appear confusing to begin with, but is very simply understood if a couple of problems are solved.

Example 5:- Approximate Magnitude Plot

$$G(s) = \frac{2500}{s(s+5)(s+50)} = \frac{2500}{s*5\left(1+\frac{s}{5}\right)*50*\left(1+\frac{s}{50}\right)}$$

Or $G(s) = \frac{10}{s(1+0.2s)(1+0.02s)}$ two corner frequency $\omega_{c1} = 5$ rad/sec, $\omega_{c2} = 50$ rad/sec.

$$G(s) = \underbrace{K_1}_{\text{Term 1}} * \overbrace{\frac{1}{s}}^{\text{Term 2}} * \underbrace{\frac{1}{1+\tau_1 s}}_{\text{Term 3}} * \overbrace{\frac{1}{1+\tau_2 s}}^{\text{Term 4}}$$

Approximation of term 3 and 4 and magnitude software is 1 and 2 are plotted on the Bode plot as shown.. The contributions of these are added "slope wise".

For example up to $\omega_{C_1} = 5$ rad/sec, only two terms have non zero contribution: K_1, s^{-1}; K_1 contribution 0 dB/ decade and s^{-1} Contributes - 20dB/ decade. So, the net slope up to $\omega_{C_1} = -20$dB/ decade. After ω_{C_1}, another term $(1+0.2s)^{-1}$ contributes another - 20 dB/ decade; afterω_{C_2}, the term $(1+0.02s)^{-1}$ contributes another - 20 dB/ decade.

Thus, the total magnitude plot is obtained as three straight line segments: (i) with slope of - 20 dB/ decade, (ii) with slope of - 40 dB/ decade, and (iii) with slope of - 60 dB/ decade. It is possible to draw (asymptotic) approximation to the phase plot also, with a maximum error of 6°.But for this example, exact phase information is used to draw phase plot as shown in the semi-log graph sheet.

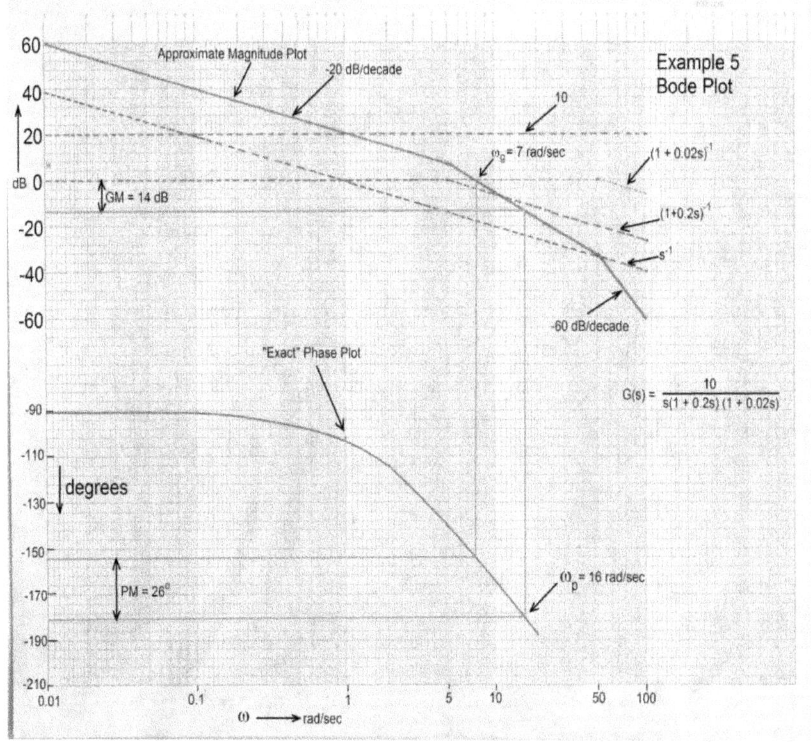

From the Bode plots, GM = 14 dB

$$PM = 31^o$$

Exact computation of Gain Margin(GM) and Phase Margin (PM) is easily done in MATLAB and is shown in the graph shown below. The manually drawn graph and the approximation for magnitude plot give slightly varying results since these involve manual errors in plotting. However, this should not form a reason for not attempting to draw Bode plot. At least three or four problems should be compulsorily practiced by students to appreciate the usefulness of the Bode Plots and also gain intuition into understanding and interpreting the MATLAB generated Bode plots.

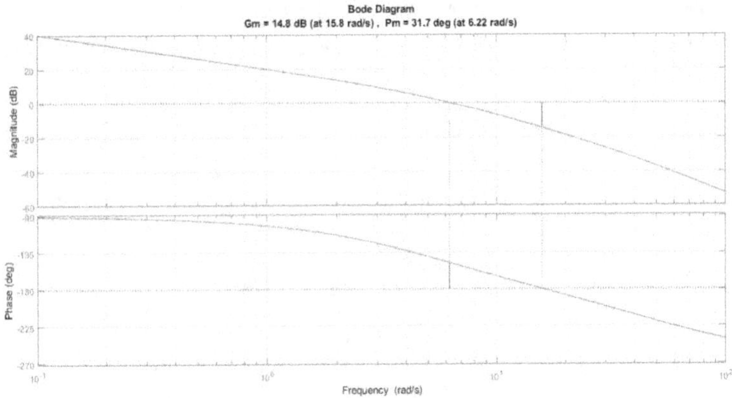

- Students are advised to carefully follow this problem and solve it on their own and convince themselves that they understood it completely.

- The approximate phase plot mentioned may be drawn as shown below for the factor of the form $(1 + \tau s)^{-1}$.

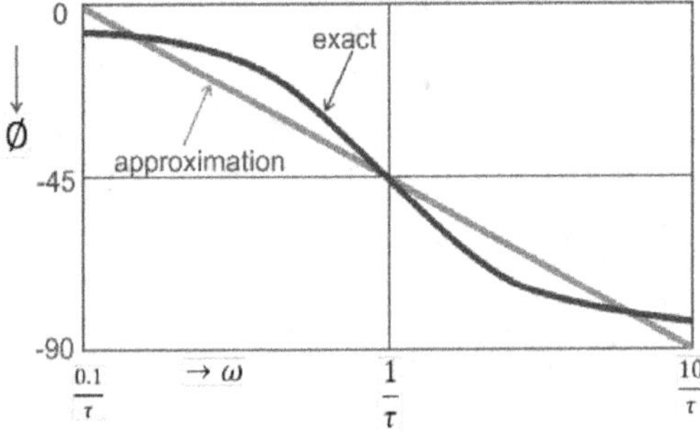

While drawing the plot, the lines were drawn very thick to ensure they come find in scanning and xerox. It is enough if they are drawn light: construction lines with H and final line in HB pencils, perhaps.

Since MATLAB can be used to obtain board diagram and GM, PM very easily, one maybe led do think that it is futile to learn how to draw Bode diagram. However, and understanding of drawing the bode diagram lets one understand and interpret computer generated Bode plots and use computer programs such as MATLAB more efficiently.

Home Work 19

1) The forward path transfer function of a Unity feedback control system is

$$G(s) = \frac{K}{s(s + 6.54)}$$

Analytically, find the resonant peak, M_r, resonant frequency ω_r, and bandwidth BW of the closed loop system when:

a) K = 5   b) k = 21.39   c) K = 100

2) The closed loop frequency response $|M(j\omega)|$.vs. frequency of a second order prototype system is shown in figure below. sketch the unit step response of the system; indicate the values of peak overshoot, peak time, and the steady state error due to a unit step input.

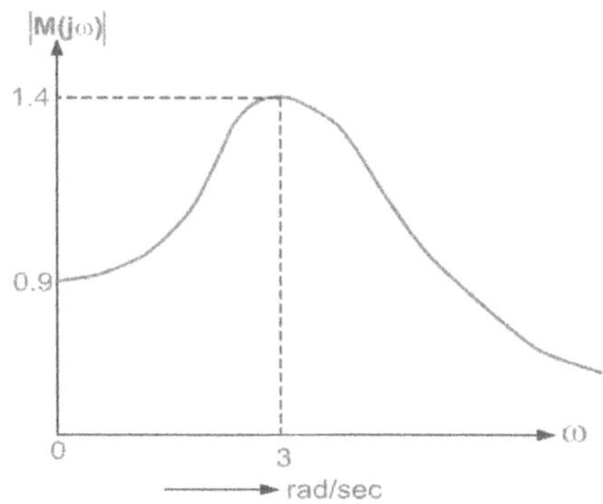

3) The forward path transfer function of a unity feedback control system is

$$G(s)H(s) = \frac{80e^{-0.1s}}{s(s + 4)(s + 10)}$$

Draw the Bode plot and find the gain margin and phase margin.

4) The forward path transfer function of a Unity feedback control system is:

$$G(s) = \frac{K(s + 1)}{s(1 + 0.1s)(1 + 0.2s)(1 + 0.5s)}$$

Plot the Bode diagram and:

1. Find the value of K so that the gain margin is 20dB.

2. Find the value of K so that the phase margin is 45°.

5) In an attempt to study the stability of a unity feedback closed loop system, experiments were conducted on the open loop system to obtain the frequency response. The plot below shows the magnitude frequency response plot.

1. From the experimental plot, compute and estimate of the transfer function.

2. From the estimated transfer function, obtain gain margin and phase margin.

3. From the GM and PM determine the maximum value of Forward path gain K for stability.

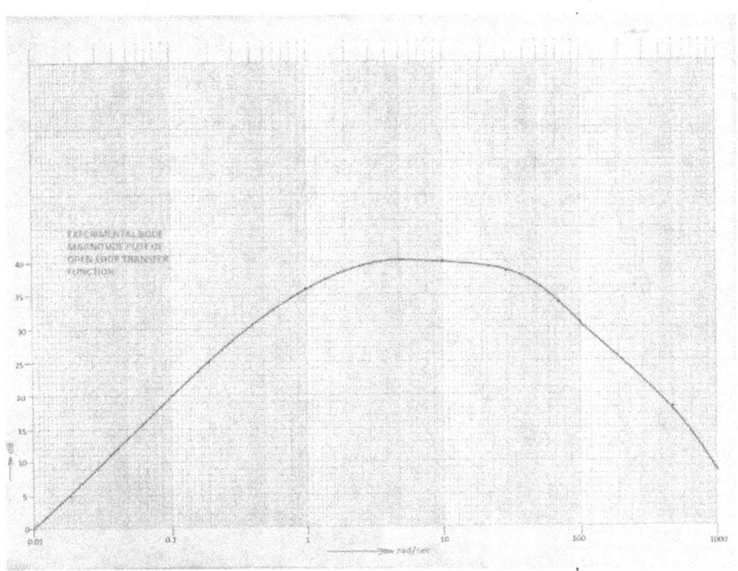

The more curious students may want to know if there is a method to obtain an estimate of the transfer function from time-domain data. Such students may search for "Generalized Partial Autocorrelation Function (GPAC)" and "Box-Jenkins Method".

1.24. Design of Compensators

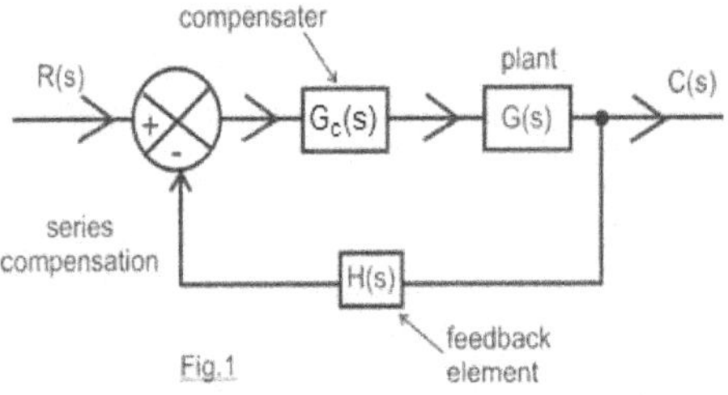

To improve the "performance" of the "system", either the plant must be changed to meet the performance specification, or the controller must be changed both may need to be changed.. Many-a-times, changing the "plant" may NOT be possible. In such case, a "compensator" is inserted into close loop system.

The compensator may be inserted in "series" or "parallel" as soon as in Figure 1 and 2.

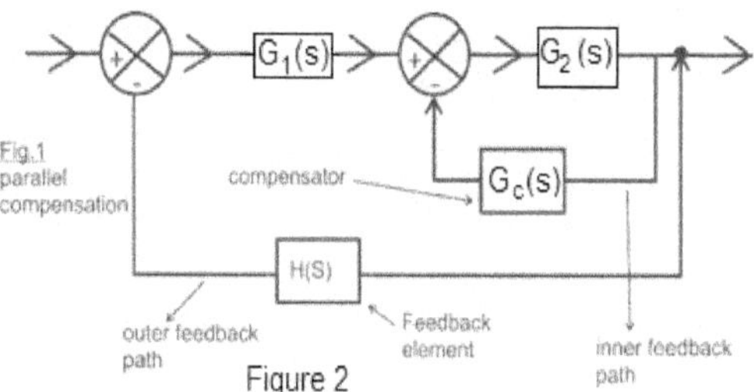

Figure 2

In series compensation, the compensator is placed in series with the plant as shown in Figure 1. In parallel compensation, the compensator is placed in the inner feedback path as shown in figure 2. Such configuration may be needed, for example in the case of inner-speed loop used in the case of position control system. The choice between series compensation and parallel compensation depends on the nature of signal in the system, power levels at various points, available components, cost of building control system, the designer's experience and so on.

In general series compensation may be easier than parallel compensation. Also, for the parallel composition scheme shown in figure 2, the number of components required may be lesser than for series compensation. In this unit, we consider the series compensation using method that use frequency response data.

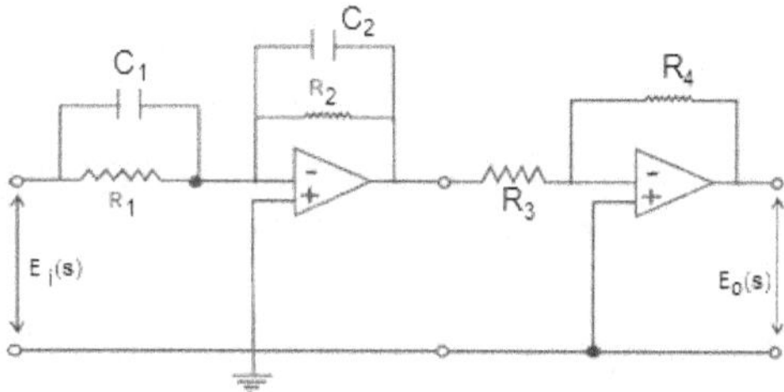

Once the mathematical design of compensator is undertaken, the designer need to come up with a device which has the required transfer function as obtained in the data is obtained in the mathematical design. Many novel Idea are proposed in Literature for constructing compensating devices. In electronic domain, circuits employing operational amplifier are commonly used. The circuit employing two sign inverting op amps as shown in figure beside as an example is often used as compensating network. Depending on the value of R_1, C_1, R_2 and C_2

this type of network or networks in series maybe used to realize a:(i) lead compensator ii) lag compensator and iii) lag-lead compensator.

Lead Compensator and Lag Compensator

The transfer function of the network shown is:

$$\frac{E_o(s)}{E_C(s)} = K_C \frac{s + \frac{1}{T}}{s + \frac{1}{\alpha T}} = K_C \alpha \frac{Ts + 1}{\alpha Ts + 1} \quad \text{where} \quad \begin{aligned} T &= R_1 C_1 \\ \alpha T &= R_2 C_2 \\ K_c &= \frac{R_4}{R_3} \frac{C_1}{C_2} \end{aligned}$$

If $\alpha < 1$ $(i.e., R_1 C_1 > R_2 C_2)$ this network is a lead network. If $\alpha > 1$, this network is a lag network.

If a sinusoidal input is applied at the input of a lead network, then the steady-state output exhibits a *phase lead.*

Similarly, if a sinusoidal input is applied at the input of a lag network, then the steady-state of output exhibits a *phase lag.*

In a *lag-lead network*, both phase lag and phase lead occur but in different frequency regions: phase lag occurs in the low frequency region and phase lead occurs in the high frequency region.

Terms such as lag compensator, lead compensator, lag-lead compensator are used to denote the type of network used.

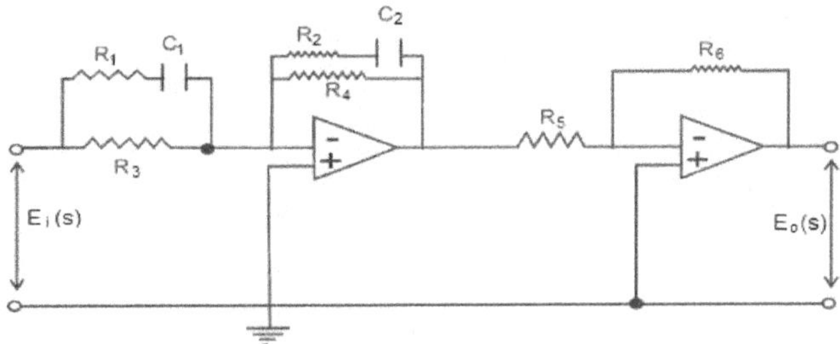

Lag-Lead Compensator

The transfer function of the above network is:

$$\frac{E_o(s)}{E_i(s)} = K_C \frac{\beta}{\gamma} \left[\frac{T_1 s + 1}{\frac{T_1}{\gamma} s + 1} \right] \left[\frac{T_2 s + 1}{\beta T_2 s + 1} \right] = K_c \frac{\left(s + \frac{1}{T_1}\right)\left(s + \frac{1}{T_2}\right)}{\left(s + \frac{\gamma}{T_1}\right)\left(s + \frac{1}{\beta T_2}\right)}$$

Where $T_1 = (R_1 + R_C)C_1, \frac{T_1}{\gamma} = R_1 C_1, \ T_2 = R_2 C_2,$ and $\beta T_2 = (R_2 + R_4)C_2.$

By properly choosing values of $R_1, R_2, R_3, R_4, R_5, R_6, C_1$ and C_2 a lag- lead network with desired characteristics may be designed.

Either root locus method or frequency response approach may be used to design a compensator.

In the root locus method, the parameters of the compensator obstruction in such a way that the root locus passes through the desired dominant poles of closed loop system to give desired transient response. The root locus method gives us direct information on the transient response of closed loop system.

In the frequency response approach, the transient response information is indirectly specified / inferred:

- Phase margin, gain margin resonant peak imply a rough estimate of damping.
- Gain crossover frequency, resonant frequency, bandwidth indicate rough estimate of speed of transient response.
- Static error constants indicates the steady state accuracy.

There are two approaches in frequency domain design (i) polar plot approach and (ii) Bode Plot approach.

We choose Bode diagram approach because:

a) Drawing closed loop bode plot of compensated system involve "addition" of compensator-Bode plot is the system-Bode plot and hence it is easier and quicker than the case of a polar plot

b) Change in the open loop gain doesn't change the shape of the Bode magnitude curve: it only shifts the curve up or down.

Before we delve into the details of the compensators and their design procedure, a few aspects of lead, lag, lag-lead compensators is not out of place here.

LEAD COMPENSATION

- Gives appreciable improvement in transient response.
- Small change in steady-state accuracy.
- May accentuate high frequency noise effects.
- Raises the order of system by 1 (Unless cancellation Raises).

LAG COMPENSATION

- Give appreciable improvement in steady-state accuracy.

- Increased transient response time.

- Suppresses effects of high-frequency noise signal.

- Raises the order of system by 1 (unless cancellation occurs).

LAG-LEAD COMPENSATION

- combines characteristics of both lead compensation and lag compensation

- raises order of system by 2 (unless cancellation occurs)

- system becomes complex

PROPER CHOICE OF COMPENSATION SCHEME MAY BE MADE ONLY UPON THROUGH STUDY OF THE SYSTEM.

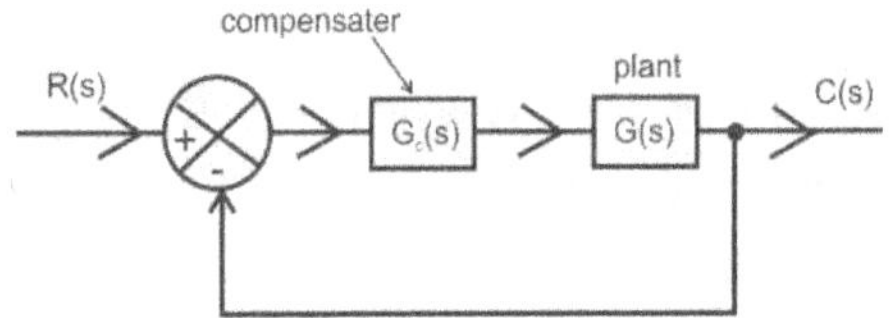

LEAD COMPENSATION TECHNIQUE

The primary function of the lead compensator shown in the figure beside is to reshape the frequency response curve to provide sufficient phase lead angle to offset the excessive phase lag associated with the components of the plant or system. We will assume that the performance specifications are given in terms of one or more of the following:

- Phase margin

- Gain margin

- Error constants

- Settling time (for dominant poles) etc.

The procedure for designing a lead compensator by the frequency response method may be stated as follows:

1) The lead compensator, $G_C(s) = K_C \alpha \dfrac{Ts+1}{\alpha Ts+1} = K_C \dfrac{s+\frac{1}{T}}{s+\frac{1}{\alpha T}}$ with $0 < \alpha < 1$. If we define $K_C \alpha = K$, we may write: $G_C(s) = K \dfrac{Ts+1}{\alpha Ts+1}$. The open – loop transfer function of compensation system is :

$$G_C(s)G(s) = K \frac{Ts+1}{\alpha Ts+1} G(s) = \frac{Ts+1}{\alpha Ts+1} G_1(s)$$

Where $G_1(s) = KG(s)$.

Derive the value of Gain K to satisfy requirement on the given static error constants.

2) Using the K determined in step 1, draw Bode diagram of $G_1(j\omega)$ and evaluate the phase margin.

3) Determine the necessary phase-lead angle [gain margin required minus the phase margin computed in step 2]. Let this be $\emptyset_m$. Typically $5°$ to $12°$ is added to $\emptyset m$.

4) Compute α from $\sin\emptyset_m = \frac{1-\alpha}{1+\alpha}$. From the Bode plot drawn, find the frequency at which magnitude is $-20log_{10}\left(\frac{1}{\sqrt{\alpha}}\right)$. Let this frequency be ω_m. Evaluate T from $G_m = \frac{1}{T\sqrt{\alpha}}$.

5) Zero of $G_C = -\frac{1}{T}$; pole of $G_C = -\frac{1}{\alpha T}$.

6) Using the value of K computed in step 1, and α computed in step 4, $K_C = \frac{K}{\alpha}$.

7) Check gain margin to see if it is satisfactory. If NOT, modify pole-zero locations of compensator until a satisfactory result is obtained.

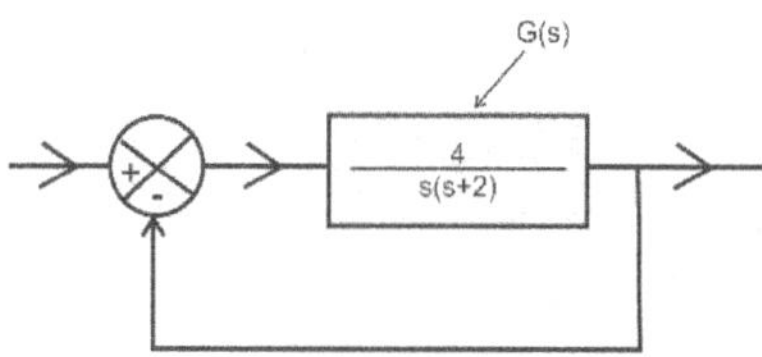

Example 1: Lead Compensator

For the system shown in figure, design a compensator for the system so that the static velocity error constant is $20\ sec^{-1}$, the phase margin is at least $50°$ and the gain margin is at least 10dB.

Solution

1) Assume the compensator of the form

$$G_C(s) = K_C \propto \frac{Ts+1}{\alpha Ts+1} = K_C \frac{s+\frac{1}{T}}{s+\frac{1}{\alpha T}}, 0 <\alpha< 1$$

If we define $K = K_C \propto, G_C(s) = K\frac{Ts+1}{\alpha Ts+1}$

Taking closed loop system to be in the form shown in figure beside, the open loop transfer function of the compensated system is

$$G_C(s)G(s) = K\frac{Ts+1}{\alpha Ts+1}G(s) = K\frac{Ts+1}{\alpha Ts+1}\cdot\frac{4}{s(s+2)} = \left(\frac{Ts+1}{\alpha Ts+1}\right)\underbrace{\left(\frac{4K}{s(s+2)}\right)}_{G_1(s)}.$$

Or, $G_C(s)G(s) = \dfrac{Ts+1}{\alpha Ts+1}G_1(s)$ where $G_1(s) = \dfrac{4K}{s(s+2)}$

Given that $K_v = \lim_{s\to 0} sG_c(s)G(s) = \lim_{s\to 0} s.\dfrac{Ts+1}{\alpha Ts+1}\dfrac{4K}{s(s+2)} = 2K = 20$.

Therefore K = 10 and $G_1(s) = \dfrac{40}{s(s+2)}$.

Think about
infinite GM!

2)   Bode plot of $G_1(s)$ is shown in Plot No. 1 and we read, $\emptyset_m = 18^o$ and $\overline{GM = \infty dB}$.

3)   We need 50° of phase margin and we have only 18°. The remaining margin of $\emptyset_m = 32^o$ must be contributed by compensator. Let $\emptyset_m = 38^o$ taking 6° more to be on a safe side.

4) $\sin\emptyset_m = \dfrac{1-\alpha}{1+\alpha} \Rightarrow \alpha = \dfrac{1-\sin 38^o}{1+\sin 38^o} \cong 0.24.$ $-20\log_{10}\left[\dfrac{1}{\sqrt{\alpha}}\right] = -6.2\text{dB}$.

As seen on Plot No.1, - 6.2dB occurs at 9 rad/sec. Take

5) $9 = \dfrac{1}{\sqrt{\alpha}T} \Rightarrow \dfrac{1}{T} = 9\sqrt{\alpha} = 4.41$ and so $\dfrac{1}{\alpha T} = \dfrac{4.49}{0.24} = 18.4$.

6) $K_c = \dfrac{K}{\alpha} = \dfrac{10}{0.24} = 41.7.$

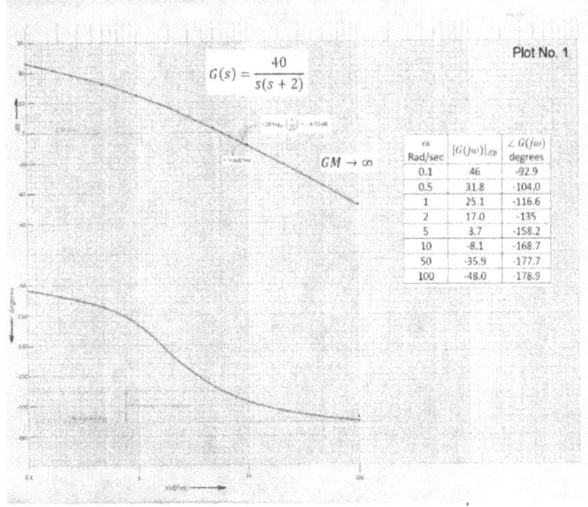

ω Rad/sec	$|G(j\omega)|_{dB}$	$\angle G(j\omega)$ degrees
0.1	46	-92.9
0.5	31.8	-104.0
1	25.1	-116.6
2	17.0	-135
5	3.7	-158.2
10	-8.1	-168.7
50	-35.9	-177.7
100	-48.0	-178.9

Substituting all the computed values, the compensated system has the open-loop transfer function.

$$G_C(s)G(s) = \frac{Ts+1}{\alpha Ts+1}\cdot\frac{40}{s(s+2)} = \frac{\left(s+\frac{1}{T}\right)T}{\left(s+\frac{1}{\alpha T}\right)\alpha T}\cdot\frac{40}{s(s+2)}$$

152

$$\text{Or } G_C(s)G(s) = \frac{166.7(s+4.41)}{s(s+2)(s+18.4)}.$$

Bode plot of the compensated open loop transfer function $G_C(s)G(s)$ is shown Plot No.2.

From the bode plot, It is seen that:

gain margin[21] $\approx 75dB$

phase margin[22] $\approx 50^o$

Thus, both the gain margin and phase margin required are met. also, the static velocity error constant for the compensated open loop transfer function is obtained as:

$$K_\vartheta = \lim_{s\to 0} sG_C(s)G(s) = \lim_{s\to 0} s\frac{166.7(s+4.41)}{s(s+2)(s+18.4)} = \frac{166.7*4.41}{2*18.4} \approx 20sec^{-1}$$

$\therefore$ Desired static velocity error constant is also achieved. Hence the design is complete and the compensator is: $G_C(s) = 41.7\frac{(s+4.41)}{(s+8.4)}$.

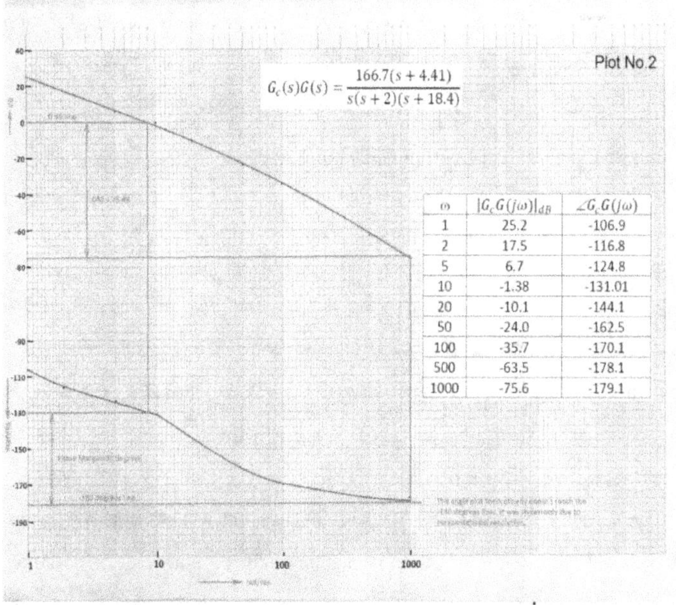

$$G_c(s)G(s) = \frac{166.7(s+4.41)}{s(s+2)(s+18.4)}$$

ω	$|G_cG(j\omega)|_{dB}$	$\angle G_cG(j\omega)$
1	25.2	-106.9
2	17.5	-116.8
5	6.7	-124.8
10	-1.38	-131.01
20	-10.1	-144.1
50	-24.0	-162.5
100	-35.7	-170.1
500	-63.5	-178.1
1000	-75.6	-179.1

---

[21] Theoretically the angle plot doesn't touch the -180° line. So, it is apt to say that $GM \to \infty$. However, as the tabulated angles show, it is coming very close to -180°.Hence GM is shown to be 75 dB. either way minimum GM is anyway achieved.

[22] Note that we took 6° more and targeted for 18° + 38° = 56° of phase margin but we obtained 50° for the compensated system. The addition of 6° more $\varnothing M$ can now be appreciated!

LAG COMPENSATION

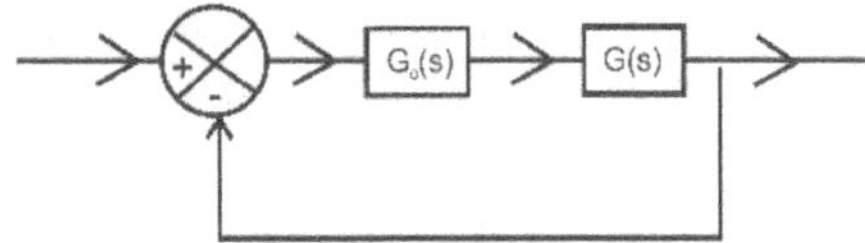

The primary function of a lag compensator is to provide attenuation in high frequency range to suppress the effects of high frequency noise. the phase-lag characteristic is of no consequence in lag compensation.

The procedure for designing lag compensator in the configuration shown besides may be listed as:

1. Choose the lag compensator of the form: $G_C(s) = K_C\beta\dfrac{(Ts+1)}{(\beta Ts+1)} = K_C\dfrac{\left(s+\frac{1}{T}\right)}{\left(s+\frac{1}{\beta T}\right)}, (\beta > 1)$.

    Define $K_C\beta = K$. Then $G_C(s) = K\dfrac{(Ts+1)}{(\beta Ts+1)}$ and the open loop transfer function of the compensated system is:

    $$G_C(s)G(s) = K\frac{(Ts+1)}{(\beta Ts+1)}G(s) = \frac{(Ts+1)}{(\beta Ts+1)}G_1(s) \text{ where } G_1(s) = KG(s)$$

    Find gain K to satisfy the requirement of the given static velocity error constant.

2. Draw Bode plot of the gain adjusted but uncompensated system $G_1(j\omega) = KG(j\omega)$. If the Specification on gain margin and phase margin are ssatisfied, well, design is complete. If   specifications are NOT satisfied, then, find the frequency (ω_1) where phase angles $=-180° + \emptyset$ where $\emptyset$ is equal to required phase margin + about 5° to 12° to ensure account for phase lag. At this frequency find gain in dB. Let this gain be x. Find β such that $20\ log_{10}\left(\frac{1}{\beta}\right) = -x$. What this does is to "slide" the gain curve down so that at frequency (ω_1) the gain curve poses through 0 dB (or ω_1 is new gain crossover frequency).

3. Choose corner frequency $\omega = \dfrac{1}{T}$ (i.e., zero of lag compensator) 1 octave or one decade below ω_1. If time constant of the lag compensator do not become too large, choose $\omega = \dfrac{1}{T}$, 1 decade below ω_1.

4. Determine the other corner frequency $\omega_C = \dfrac{1}{\beta T}$. From the value of T obtained in (3) and β obtained from Step (2).

5. $K_C = \dfrac{K}{\beta}$ where K is obtained in step (1) and β is obtained in step (2)

6. It is a good idea to draw bode plot of $Gc(s)G(s)$ and check GM and $\emptyset M$.

Example 2: Lag Compensator

Open loop transfer function of a system is given by

$$G(s) = \frac{1}{s(s+1)(0.5s+1)}$$

It is desired that high frequency noise effects are to be suppressed by choosing a lag compensator. Design the lag compensator to achieve static velocity error constant $K_v = 5$ sec^{-1}, phase margin of at least 40°, and the gain margin of at least 10 dB.

Solution

1) The leg compensator is taken as: $G_C(s) = \overset{K}{\overbrace{K_C \beta}} \frac{(Ts+1)}{(\beta Ts+1)} = K_C \frac{\left(s+\frac{1}{T}\right)}{\left(s+\frac{1}{\beta T}\right)}, (\beta > 1)$

   Since $K_v = sec^{-1}$ is desired, we have

$$K_v = \lim_{s\to 0} sG_C(s)G(s) = \lim_{s\to 0} sK \frac{(Ts+1)}{(\beta Ts+1)} \cdot \frac{1}{s(s+1)(0.5s+1)} = K = 5.$$

With K = 5, the gain-adjusted, but uncompensated transfer function G₁(s) may be written as:

$$G_C(s) = \frac{5}{s(s+1)(0.5s+1)}$$

2) The Bode[23] plot of G₁(s) is shown in Plot No. 3. It is clearly seen that the system is unstable. We need 40° phase margin; with additional 12°, the required phase margin aspired for is 52°.

3) With this, phase angle is -(180-52) or -128°. On the phase curve, -128° occurs at $\omega_1 = 0.5$ rad/sec. Take the corner frequency corresponding to zero of the lag compensator to be 0.1 rad/sec$= \frac{1}{T} \Rightarrow T = 10$.

4) Further, at 0.5 rad/sec, $|G_1(j\omega)| = 20$dB. If we want to make $\omega_1 = 0.5$ as new gain crossover frequency, the lag compensator must have an attenuation of -20dB$\Rightarrow$ $20\,log_{10}\left(\frac{1}{\beta}\right) = -20 \Rightarrow \beta = 10$. Therefore the other corner frequency corresponding to the pole of lag compensator is $\frac{1}{\beta T} = 0.01$rad/sec.

5) $K_C = \frac{K}{\beta} = \frac{5}{10} = 0.5$

   ∴ Open loop transfer function of compensated system is obtained as:

---

[23] Since several examples of drawing Bode Plots manually are shown, from now on, MATLAB will be used to general the Bode Plots.

$$Gc(s)G(s) = \frac{5(10s + 1)}{s(100s + 1)(s + 1)(0.5s + 1)}$$

Bode plot shown in Plot No.4 shows that the system has desired phase margin and gain margin. Hence design is complete and $G_c(s) = 0.5\,\dfrac{s+\frac{1}{10}}{s+\frac{1}{100}}$.

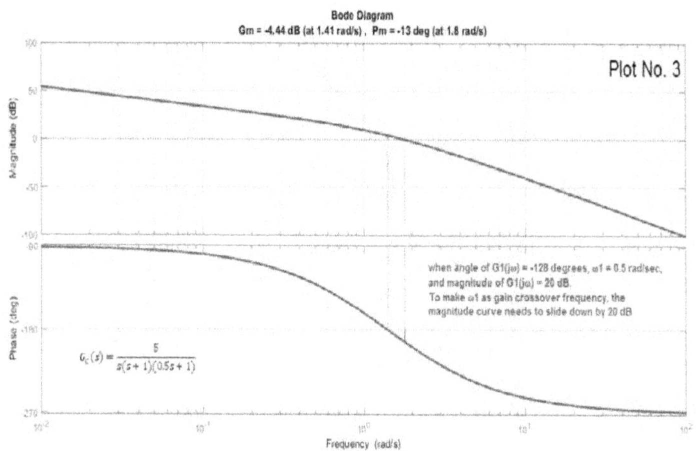

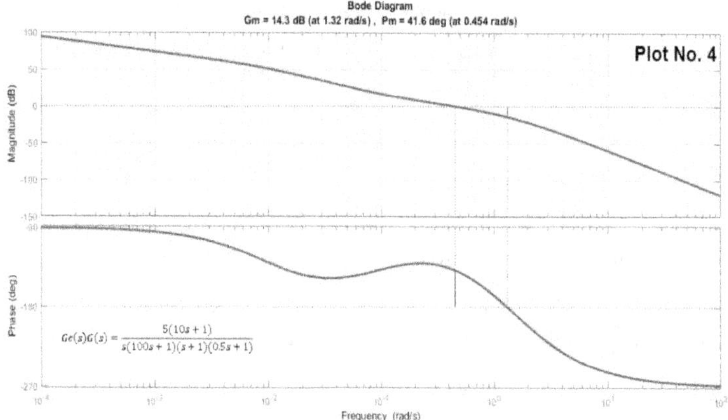

LAG-LEAD COMPENSATION

The lag lead compensator is taken as:

$$G_C(s) = K_C \frac{\left(s+\frac{1}{T_1}\right)\left(s+\frac{1}{T_2}\right)}{\left(s+\frac{\beta}{T_1}\right)\left(s+\frac{1}{\beta T_2}\right)} = K_C \frac{(T_1 s+1)(T_2 s+1)}{\left(\frac{T_1}{\beta}s+1\right)(\beta T_2 s+1)} \text{ with } \beta > 1$$

Terms involving T_1 form phase - lead portion of compensator and the terms involving T_2 for phase-lag portion.

The phase-lead portion adds phase-lead angle and increases phase margin. The phase lag portion provides attenuation near and above gain crossover frequency to improve steady state performance.

Example 3: Lag-Lead Compensation

Consider a Unity feedback system with open-loop transfer function.

$$G(s) = \frac{K}{s(s+1)(s+2)}$$

Design a Lag-lead compensator to achieve: (i) state velocity error constant of 10 sec⁻¹, (ii) phase margin to be 50°, and (iii) gain margin of 10 dB or more.

Solution: Since variable gain K is available in system, assume the compensator of the form

$$G_C(s) = \frac{\left(s+\frac{1}{T_1}\right)\left(s+\frac{1}{T_2}\right)}{\left(s+\frac{\beta}{T_1}\right)\left(s+\frac{1}{\beta T_2}\right)} \quad \text{i.e., } K_C = 1.$$

$$\text{Given } K_v = 10\ sec^{-1} = \lim_{s \to 0} sG_C(s)G(s) = sG_C(s)\frac{K}{s(s+1)(s+2)} = \frac{K}{2}$$

Therefore K = 20. Draw bode plot of $G(s)$ with $K = 20$ as shown in Plot No.5. Since GM and $\emptyset M$ are negative, the system is unstable. Let us choose used to have 50° of phase angle at 1.5 rad/sec and choose this as gain crossover frequency. Let us choose the zero of phase lag portion of network $\frac{1}{T_2}$ one decade below gain crossover frequency: $\frac{1}{T_2} = 0.15$ rad/sec. We will find β in the following.

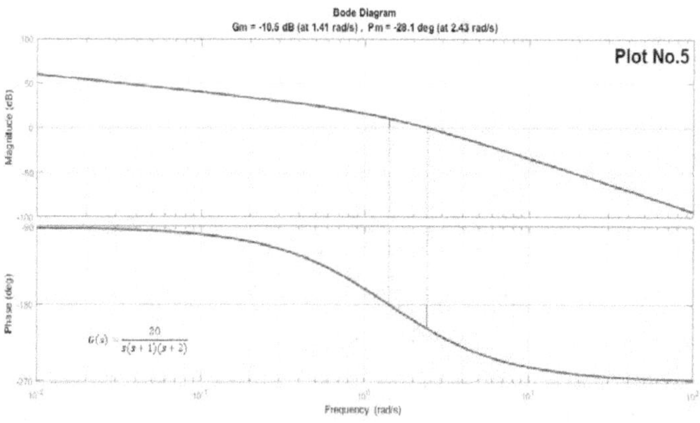

We need $\emptyset_m$ of 50°. Take a bit more, 55° and use the relation:

$$\sin \emptyset_m = \frac{1-\frac{1}{\beta}}{1+\frac{1}{\beta}} = \frac{\beta-1}{\beta+1} = \sin 55^o$$

to obtain $\beta \approx 10$. With this β, the corner frequency $\frac{1}{\beta T_2}$ (corresponding to the pole of phase lag portion of the compensator) will be 0.015 rad/sec. Hence, the transfer function of the phase-lag portion of the lag lead compensator becomes:

$$\frac{s+0.15}{s+0.015} = 10 \, \frac{(6.67s+1)}{(66.67s+1)}$$

Choosing 0.7 rad/sec and 7 rad/sec (to maintain $\beta = 10$) as corner frequencies of the TF of lead portion, We get the TF of lead portion as

$$\frac{s+0.7}{s+7}.$$

With this, the TF of the compensator and plant in open-loop is

$$G(s)G_C(s) = \left(\overbrace{\frac{(s+0.7)}{(s+7)}\frac{(s+0.15)}{(s+0.015)}}^{\text{Compensator}}\right)\left(\overbrace{\frac{20}{s(s+1)(s+2)}}^{\text{System}}\right)$$

Bode plot of $G(s)G_C(s)$ is shown in Plot No. 6. It is seen that $GM, \emptyset M$ requirements are met. Also, $K_v = 10sec^{-1}$ is also met. Hence the compensator design is complete.

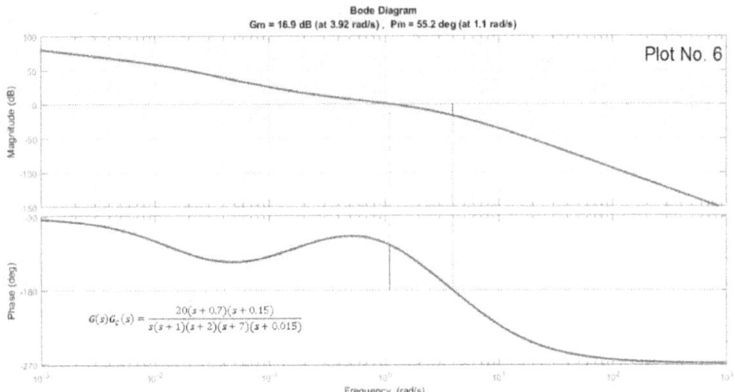

$$G(s)G_c(s) = \frac{20(s+0.7)(s+0.15)}{s(s+1)(s+2)(s+7)(s+0.015)}$$

Students, and many a practitioner, may feel that drawing bode plots by hand is tedious and unnecessary especially since special packages such as MATLAB draw Bode plot at the click of a mouse, so to say. While this is true, sketching Bode plot by hand (either accurate/exact and / or asymptotic approximation) gives a deeper understanding and appreciation of the science of control engineer.

Home Work 20

1) Solve each of the examples presented on your own without looking into the solution. Refer to the solution only when you are stuck at a point for more than 20 minutes. Keep trying till you can solve the entire problem completely on your own. This process takes maximum of 4 hours.

The following objective type questions are provided for the sake of practice. Go through each one carefully. Check your answers with your instructor or on the text-book web-site.

1	Does a system with closed loop mathematical block diagram always have an external sensor built into it?
A	Sometimes it does and sometimes it doesn't.
B	No, it will never have
C	Yes, it will always have.
D	It is a stupid question.
2	A system is defined by the equation $y = m * x + c$ where x is the input to the system, y is the output from the system, and m,c are non-zero real constants. Is this system a linear system?
A	Yes it is a linear system because it is a straight line.
B	It is linear only when m and c are positive.
C	No it is not a linear system.
D	It is not possible to propose an answer to this question.
3	A system is defined by the equation $y(t) = x(t+1)$ where $y(t)$ is output from the system, $x(t)$ is input to the system, and t is time. Is it causal system?
A	When x is a continuous function of time, it is causal.
B	When y is a continuous function of time, it is causal.
C	When x and y are continuous, it is causal.
D	It is non-causal system.
4	Is a nonanticipative system a causal system?
A	Some nonanticipative systems are causal systems.
B	Some causal systems are nonanticipative systems.
C	All causal systems may be termed nonanticipative systems.
D	It is not possible to provide a definite answer to this question.
5	Let us define a function $u_\tau : R \to R^m$ given by

	$u_\tau(t) = \begin{cases} u(t) \text{ for } t \leq \tau \\ 0 \text{ for } t \geq \tau \end{cases}$. Similarly define a function $y_\tau : R \to R^p$. Further a system is defined by the mapping $y = T(u)$. It is known that $\forall \tau \in R \wedge \forall u, (T(u))_\tau = (T(u_\tau))_\tau$. Then the mapping given by $y = T(u)$ is
A	A deterministic system
B	A stochastic system
C	A non-inverting symbiotic system
D	A causal system
6	Let us define a function $u_\tau : R \to R^m$ given by $u_\tau(t) = \begin{cases} u(t) \, for \, t \leq \tau \\ 0 \, for \, t \geq \tau \end{cases}$. Similarly define a function $v_\tau : R \to R^m$. Let us further consider a mapping $T : R^m \to R^p$ to define a system given by $y = T(u)$. Suppose it is known that $u_\tau = v_\tau \Rightarrow (T(u))_\tau = (T(v))_\tau \, \forall \tau \in R$. Then the mapping T may be considered as
A	A deterministic system
B	A stochastic system
C	A non-inverting symbiotic system
D	A causal system
7	A continuous time system is defined by the equation $y(t) = u(t - 1)$. What type of system is it?
A	Purely imaginary but stochastic system.
B	A Discrete time system.
C	A symbiotic, non-causal, unpredictable system.
D	A system with memory.
8	Let Z be a set of positive integers and a discrete time system with input x(k) at sampling instant $k \in Z$ and output y(k) at sampling instant $k \in Z$ be defined as $$y(n) = \sum_{k=-\infty}^{n} x(k) \text{ for } n, k \in Z$$ Then, the system with x(k) as the input and y(k) as output may be classified as:
A	An infinite memory stochastic, discrete-time system with no loss.
B	A non-deterministic, non-causal, dynamic system with memory.
C	A deterministic, causal, discrete-time, dynamic system with memory.
D	Similar to an amnesia patient.

9	Let R be the set of real numbers and $x, y: R \to R$ be the input and output of a system defined by $$y(t) = \int_{-\infty}^{t} x(\tau)d\tau$$ where t is the time in seconds. Then the system may be classified as
A	Infinite dimensional, scalar, deterministic, dynamic system with infinite memory.
B	Finite dimensional, scalar, deterministic, continuous-time dynamic system.
C	A non-causal system
D	A stochastic system
10	Let R be the set of real numbers and let $\alpha \in R$. Introduce shift operator $Q_\alpha: R \to R$ as $$Q_\alpha u(t) = u(t - \alpha), u, t \in R.$$ Further, a system that is represented by the mapping y=T(u) is such that $$TQ_\alpha(u) = Q_\alpha(T(u)) = Q_\alpha(y).$$ Then the system is classified as
A	Stochastic, discrete time, α-time gap dynamic system.
B	Deterministic, discrete time, α-time gap dynamic system
C	It is not possible to specify the type of the system.
D	It is a time-invariant system.
11	A system is such that if the input signal is shifted by a finite amount of time, the output is also shifted by the same amount. What can you say about the system?
A	It is definitely a time-invariant system.
B	It is definitely a time-variant system.
C	It is definitely a discrete-time system.
D	The information provided is very vague and no conclusion whatsoever may be inferred from it.
12	A system defined by $y(t) = \cos u(t)$ is
A	Fuzzy system.
B	Sinusoidal system.
C	Non-causal type-1 system.
D	Time-invariant system.
13	Let Z be the set of positive integers and let $n \in Z$. A system is defined by mapping input to output as $$y(n) = nu(n).$$

	Then, the system is
A	Multiplicative, discrete time, non-causal system.
B	Stochastic, discrete time, causal system.
C	Time-variant, discrete-time, deterministic system.
D	This system cannot be classified.
14	Let Z be the set of positive integers and let $n \in Z$. Let the sequence p(n) be defined as: $$p(n) = \begin{cases} 1, \text{ for } n \geq 0, n \in Z \\ 0, \text{ for } n < 0, n \in Z \end{cases}$$ and another sequence is defined as $$\beta(n) = p(n) - p(n-1).$$ The function β is commonly known as
A	β-function.
B	Unit step input.
C	Discrete-time impulse or unit pulse or unit sample.
D	An arbitrary function without any known name.
15	Let Z be the set of positive integers. Define the functions $$\delta(n) = \begin{cases} 0, \text{ for } n \neq 0, n \in Z \\ 1, \text{ for } n = 0 \end{cases}$$ and $$p(n) = \begin{cases} 1, \text{ for } n \geq 0, n \in Z \\ 0, \text{ for } n < 0, n \in Z \end{cases}$$ Then, which of the following equations is true?
A	$$p(n) = \begin{cases} \displaystyle\sum_{k=-\infty}^{\infty} \delta(n-k), \text{ for } n \geq 0 \\ 0, \text{ for } n < 0 \end{cases}$$
B	$$p(n) = \begin{cases} \displaystyle\sum_{k=0}^{\infty} \delta(k-n), \text{ for } n \geq 0 \\ 0, \text{ for } n < 0 \end{cases}$$
C	$$p(n) = \begin{cases} \displaystyle\sum_{k=-\infty}^{\infty} \delta(k-n), \text{ for } n \geq 0 \\ 0, \text{ for } n < 0 \end{cases}$$
D	$$p(n) = \begin{cases} \displaystyle\sum_{k=0}^{\infty} \delta(n-k), \text{ for } n \geq 0 \\ 0, \text{ for } n < 0 \end{cases}$$
16	Which of the following circuits represent a memoryless system?

A	i(t) current source R_1 v(t) voltage across resistor
B	i(t) current source R_1 v(t) C_1 voltage across resistor
C	i(t) current source C_1 v(t) voltage across capacitor.
D	None of the above
17	Which of the following statements accurately describes the characteristic of feedback in control systems?
A	Feedback may be unnecessary in cases where we do not need accuracy.
B	Feedback can increase the gain of a system in one frequency range but can decrease it in another.
C	Feedback is always beneficial because it increases the gain of the system making it sensitive.
D	None of these is true.
18	Which of the following statements is a more accurate description about the closed-loop control systems?
A	Closed loop systems involve costly equipment and hence are never preferred unless the system is very essential one.
B	If there is no disturbance, then open loop system is always the faster system.
C	Closed loop system can increase the bandwidth and is usually more accurate than open-loop system.
D	None of the above.
19	"If an open-loop system is unstable, then applying negative feedback will always improve its stability." Which statement below comments more appropriately about the previous sentence.

A	It is not possible to give a generalised comment.
B	Always stability margin decreases.
C	True in all situations.
D	False However, properly designed compensation networks can allow negative feedback to stabilize a system that is unstable without negative feedback.
20	Is it true that nonlinear elements are sometimes intentionally introduced into a control system to improve its performance?
A	This is never true. We always want to have linear control systems.
B	True. For example, a limiter can prevent saturation of the electronics.
C	Even though we do not introduce nonlinear elements, all control systems are nonlinear.
D	It is not possible to give an answer to this question.
21	Is it true that discrete-data control systems are more susceptible to noise due to the nature of their signals?
A	Yes. Discrete-data systems are very susceptible to noise.
B	Some times they are susceptible in certain frequency range and in ohter situations they are not.
C	False. Discrete Data (Digital) systems have high noise immunity.
D	This question is improperly posed.
22	Is the mechanism of control of body temperature a non-feedback system? Pick most appropriate answer.
A	Yes, only when the person is suffering from Ebola fever.
B	No, it is a non-feedback system.
C	It is feedback system as the temperature of our body is regulated periodically and being warm blooded we regulate our body temperature w.r.t. to the climate and surroundings.
D	Sometimes body uses feedback and sometimes it doesn't
23	What is the effect of feedback on sensitivity of the system?
A	For a good control system the sensitivity must be less and in closed loop control system it gets reduced by the factor of $1/(1+GH)$.
B	Sensitivity is always reduced by using open-loop system.
C	Both A and B are true.
D	Both A and B are false.
24	The closed system has higher ______ than open loop control system, this implies increased speed of response.

A	Gain
B	Acceleration
C	Bandwidth
D	Frequency
25	The impulse response of a LTI system is a unit step function, then the corresponding transfer function is
A	$\dfrac{1}{s}$
B	$\dfrac{1}{s^2}$
C	1
D	s
26	For a type one system, the steady–state error due to step input is equal to
A	Infinity.
B	0.25
C	Zero
D	1
27	For a type zero system, the steady – state error due to ramp input is equal to
A	Infinity.
B	0.25
C	Zero
D	1
28	What is the order of closed loop system for the plant transfer function $G(s) = \dfrac{K}{s^2(1+\tau s)}$ and with unity feedback
A	3
B	2
C	1
D	0
29	The impulse response of a closed loop system is $c(t) = -te^{-t} + 2e^{-t}$. Its transfer function is
A	$\dfrac{2s+2}{(s+1)^2}$
B	$\dfrac{2s+1}{(s+1)^2}$

C	$$\dfrac{2s}{(s+1)^2}$$
D	$$\dfrac{2s+1}{(s+1)^3}$$
30	What is the open loop DC gain of a unity negative feedback control system having closed loop transfer function $\dfrac{(s+4)}{(s^2+7s+13)}$?
A	$$\dfrac{9}{4}$$
B	$$\dfrac{4}{13}$$
C	$$\dfrac{4}{9}$$
D	$$\dfrac{13}{9}$$
31	A certain control system has input u(t) and output c(t). If the input is first passed through a block having transfer function e^{-s} and the applied to the system. The modified output will be
A	$c(t-1) \cdot u(t-1)$
B	$c(t-1) \cdot u(t)$
C	$c(t) \cdot u(t-1)$
D	$c(t) \cdot u(t)$
32	A second order control system is defined by the following equation: $$4\dfrac{d^2 c(t)}{dt^2} + 8\dfrac{d\,c(t)}{dt} + 16c(t) = 16u(t).$$ The damping ratio and natural frequency for this system are respectively
A	0.25 and 2 rad/s
B	0.25 and 4 rad/s
C	0.50 and 2 rad/s
D	0.50 and 4 rad/s
33	Differentiation of parabolic response is a __ response?
A	Impulse
B	Parabolic
C	Step
D	Ramp

34	The impulse response of a unity negative feedback closed loop system is $c(t) = -te^{-t} + 2e^{-t}$. Its open loop transfer function is
A	$$\frac{2s + 2}{(s + 1)^2}$$
B	$$\frac{2s + 1}{(s + 1)^2}$$
C	$$\frac{2s + 1}{s^2}$$
D	$$\frac{2s + 2}{s^2}$$
35	The output of the feedback control system must be a function of
A	Reference input
B	Reference output
C	Input and feedback signal
D	Output and feedback signal
36	If x(t) and y(t) are input and output for a system and they are related by the equation y(t)=sin (x(t)), then the system is
A	Linear, time-invariant system
B	Nonlinear, time-invariant system
C	Nonlinear, time-variant system
D	Linear, time-variant system
37	A linear system at rest is subject to an input signal r(t)=1-e^{-t}. The response of the system for t>0 is given by c(t)=1-e^{-2t}. The transfer function of the system is:
A	$$\frac{(s + 1)}{(s + 2)}$$
B	$$\frac{2(s + 1)}{(s + 2)}$$
C	$$\frac{2(s + 2)}{(s + 1)}$$
D	$$\frac{2(s + 1)}{(s + 1)}$$

38	Figure below shows schematic of a system. Pick the sentence that best describes the system.
A	The field current I_f and the voltage e_b remain constant throughout the operation of the system whereas e_a, T_m, and ω_m vary throughout the operation of the system.
B	Field current I_f is kept constant and e_a is varied to obtain desired T_m and ω_m. The potential difference e_b varies as the speed of the motor varies. The steady-state torque-speed characteristics of this system is a straight line.
C	Both the above statements are true.
D	Both A and B are false.
39	The figure below shows the schematic of a system. Pick the statement that is most accurate.
A	This sketch represents Armature Controlled DC Servomotor. However, there is a mistake in the sketch: inductor L_a should not be shown because in DC circuits inductance will be absent.
B	This sketch represents a closed loop system. The mathematical block diagram also is a closed loop diagram.
C	This sketch represents an open-loop system viz., Armature Controlled DC Servomotor. However, the mathematical block diagram has a loop in it. This loop is called inherent feedback loop.
D	All the statements above are correct.

40	Figure below shows schematic of a position control system using Armature Controlled DC Servomotor. If the differential potentiometer gain is K_p , the amplifier gain is K_a , motor constant is K_m , and the back emf constant is K_b , the transfer function from θ_m to θ_r is:
	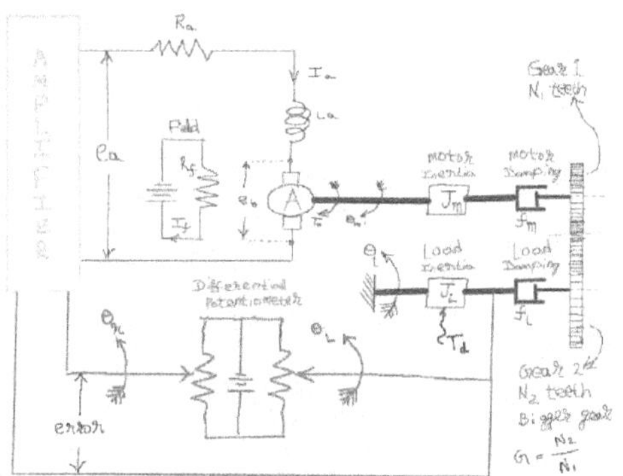
A	$$\dfrac{GK_m K_p K_a}{G\big[(R_a + L_a s)(J_{eq}s^2 + f_{eq}s) + K_m K_b s\big] + K_m K_p K_a s}$$
B	$$\dfrac{GK_m K_p K_a}{\big[(R_a + L_a s)(J_{eq}s^2 + f_{eq}s) + K_m K_b s\big] + K_m K_p K_a}$$
C	$$\dfrac{K_m K_p K_a}{G\big[(R_a + L_a s)(J_{eq}s^2 + f_{eq}s) + K_m K_b s\big] + K_m K_p K_a}$$
D	$$\dfrac{GK_m K_p K_a}{G\big[(R_a + L_a s)(J_{eq}s^2 + f_{eq}s) + K_m K_b s\big] + K_m K_p K_a}$$
41	Figure below shows schematic of an Armature Controlled DC Servomotor. If K_m is the motor constant, K_b is the back emf constant, and η is energy conversion efficiency of the motor, then pick which relation is correct:
A	$$K_m = \eta K_b$$
B	$$K_b = \eta K_m$$

C	$K_m = K_b$
D	$K_b = \eta^2 K_m$
42	Figure below shows schematic of a position control system using Armature Controlled DC Servomotor. Pick the statement that best applies to the system.
A	The schematic represents a position control system.
B	There will be two loops in the mathematical block diagram representing this system.
C	The system has a sensor in it and the system is a feedback control system.
D	All the above are accurate statements about the system.
E	Only C is true.
43	Figure below shows schematic of a system. Pick the statement that best applies to the system.
A	The system is an open-loop servomotor.
B	The system is Field Controlled DC Servomotor where armature current is maintained constant.
C	Both e_a and e_b need to be varied to maintain desired speed or torque.
D	All the above statements are true

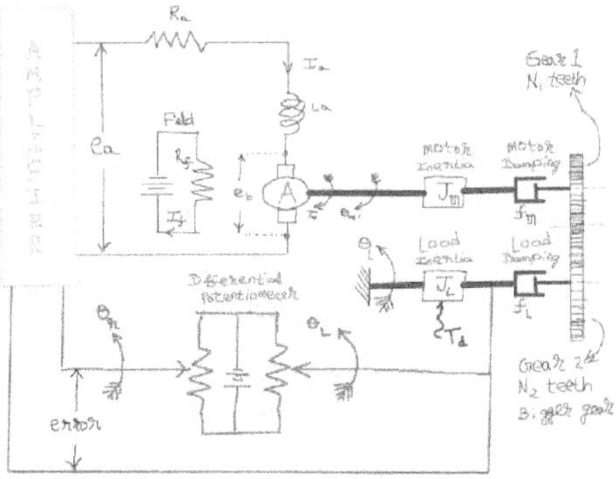

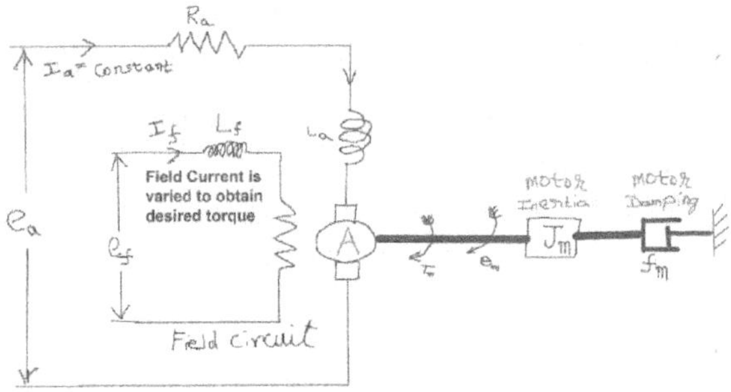

44	Figure below shows a spring-mass-damper system. Pick the sentence that is most appropriate. 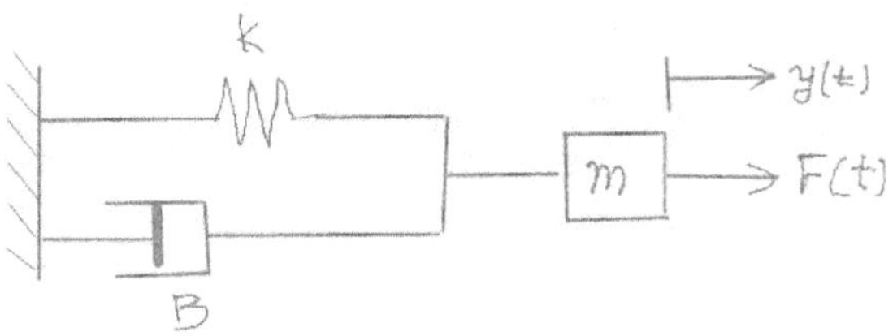
A	Spring K and damper B are in parallel.
B	Spring K, damper B are in parallel and mass m is in parallel with the combination of K and B.
C	The transfer function of the system is $$\frac{Y(s)}{F(s)} = \frac{1}{ms^2 + Bs + K}$$
D	All the above statements are true.
45	Look at the schematic shown in Figure below. Pick the statement that is most appropriate. 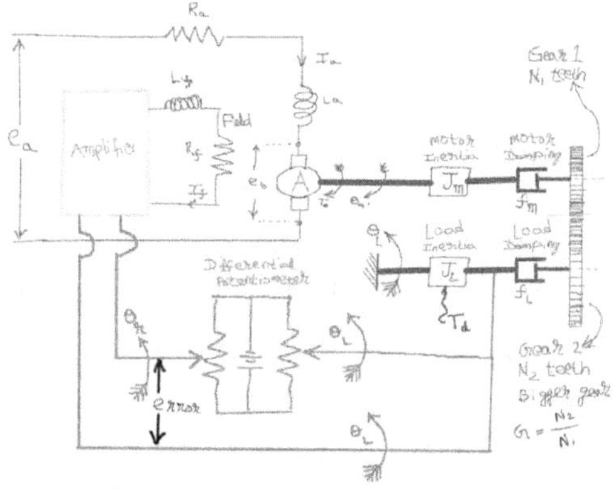
A	This figure is the schematic of an AC Servomotor used in position control application.
B	This is not a closed-loop system.
C	The net damping present in this system is much more than the corresponding system employing ACDC Servomotor.
D	None of the above statements is true
46	Pick the statement that is most appropriately true.
A	Error is always zero in open loop systems.

B	Open loop systems are always desirable.
C	Feedback is very expensive and must be avoided as much as possible.
D	Feedback is desirable in situations where, by appropriate control algorithm, we can eliminate errors without change to the plant.
47	For same motor inertia, same load inertia, same motor damping, same load damping, same currents, and same disturbances, which of the statements is most appropriate for open loop operation?
A	ACDC Servomotor has smaller time constant than FCDC servomotor and hence ACDC Servomotor is faster.
B	ACDC Servomotor has better damping characteristics and hence less oscillatory.
C	FCDC Servomotor requires bigger power supply than ACDC Servomotor.
D	All the above statements are more or less correct.
48	Figure below shows the schematic of a FCDC Servomotor. Why is inductance shown in Field circuit of FCDC Servomotor and not shown in ACDC Servomotor.
A	Because inductance is inherently present in ACDC Servomotor and inherently absent in FCDC Servomotor.
B	Because ACDC Servomotor acts as a alternating current circuit and FCDC Servomotor acts as a direct current circuit.
C	Because we can neglect inductance in both ACDC and FCDC but we prefer to retain the inductance in FCDC for simplicity.
D	None of these is an appropriate response to the question.
49	The transfer function $\frac{E_0(s)}{E_i(s)}$ for the cascaded circuit shown below is:

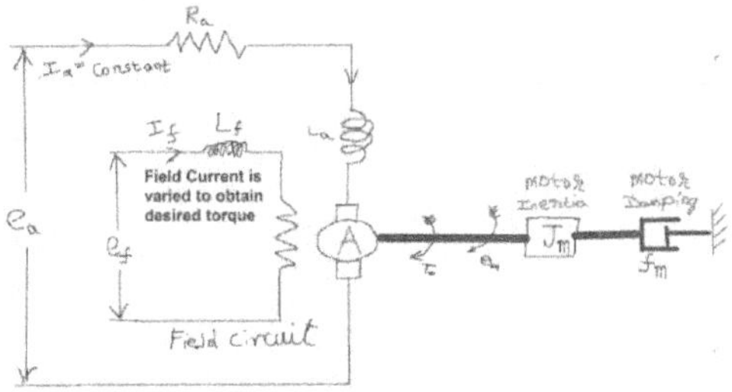

	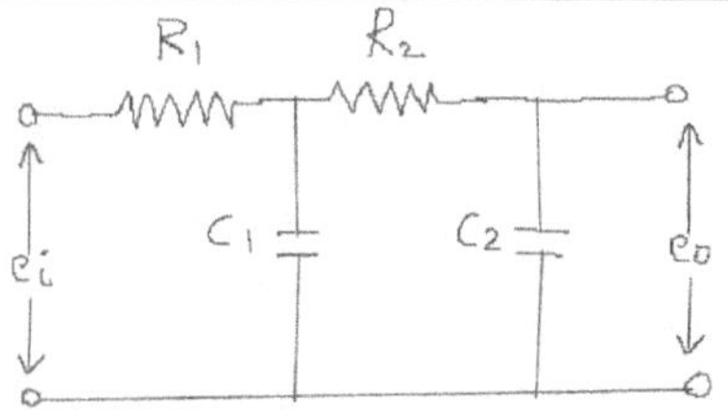
A	$\frac{1}{R_1C_1R_2C_2s^2 + (R_1C_1 + R_2C_2 + R_1C_2)s + 1}$
B	$\frac{R_1C_2s + 1}{R_1C_1R_2C_2s^2 + (R_1C_1 + R_2C_2 + R_1C_2)s + 1}$
C	$\frac{1}{R_1C_1R_2C_2s^2 + (R_1C_1 + R_2C_2 + R_2C_1)s + 1}$
D	$\frac{R_1C_2s + 1}{R_1C_1R_2C_2s^2 + (R_1C_1 + R_2C_2 + R_2C_1)s + 1}$
E	
50	The mechanical system shown in Figure (a) and the electrical system shown in Figure (b) will be analogous if: 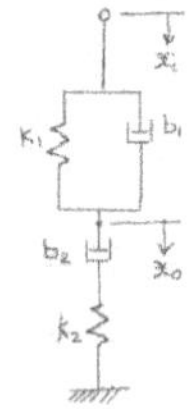 Figure (a) 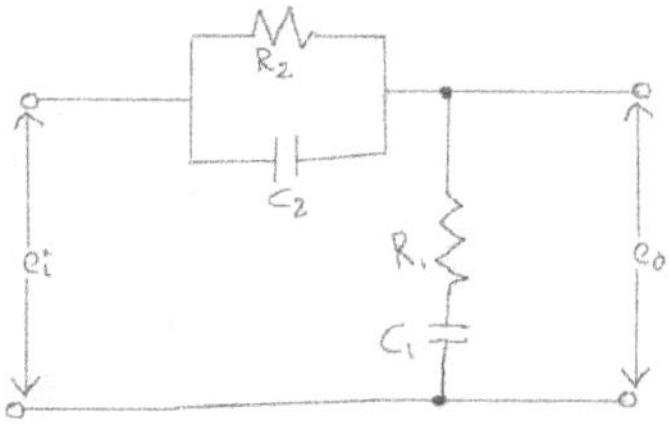 Figure (b)
A	$K_1 = C_1, K_2 = C_2, b_1 = R_1$, and $b_2 = R_2$
B	$K_1 = \frac{1}{C_1}, K_2 = \frac{1}{C_2}, b_1 = R_1$, and $b_2 = R_2$
C	$K_1 = \frac{1}{C_1} = K_2 = \frac{1}{C_2}, b_1 = R_1 = b_2 = R_2$
D	None of the above
51	Figure below shows schematic of an Armature Controlled DC Servomotor used in a position control system. The effective inertia and viscous damping as referred to the motor side may be written as:

A	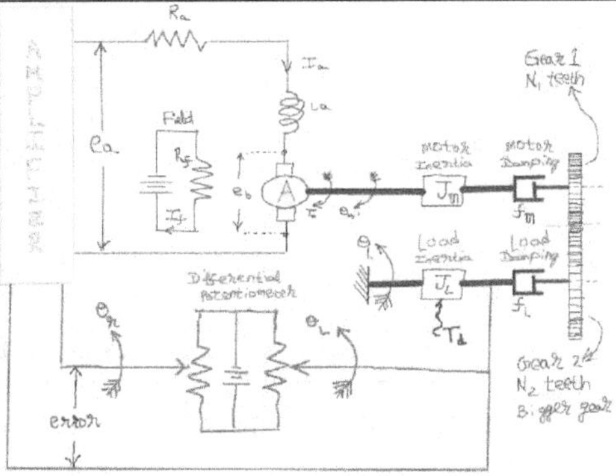
	$$J_{eq} = J_L + \left(\frac{N_2}{N_1}\right)^2 J_m \, ,$$ $$f_{eq} = f_L + \left(\frac{N_2}{N_1}\right)^2 f_m \, .$$
B	$$J_{eq} = J_L + \left(\frac{N_1}{N_2}\right)^2 J_m \, ,$$ $$f_{eq} = f_L + \left(\frac{N_1}{N_2}\right)^2 f_m \, .$$
C	$$J_{eq} = J_m + \left(\frac{N_2}{N_1}\right)^2 J_L \, ,$$ $$f_{eq} = f_m + \left(\frac{N_2}{N_1}\right)^2 f_L \, .$$
D	$$J_{eq} = J_m + \left(\frac{N_1}{N_2}\right)^2 J_L \, ,$$ $$f_{eq} = f_m + \left(\frac{N_1}{N_2}\right)^2 f_L \, .$$
52	If the transfer function $\dfrac{X_o(s)}{X_i(s)} = \dfrac{\left(\frac{b_1}{k_1}s+1\right)\left(\frac{b_2}{k_2}s+1\right)}{\left(\frac{b_1}{k_1}s+1\right)\left(\frac{b_2}{k_2}s+1\right)+\frac{b_2}{k_1}s}$, then the transfer function $\dfrac{E_o(s)}{E_i(s)}$ is:

	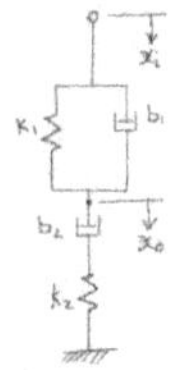 Figure (a) 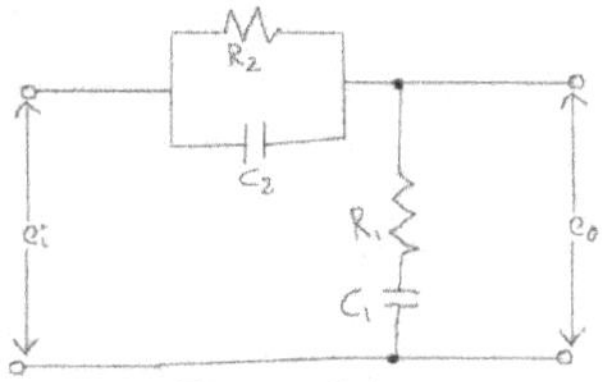 Figure (b)	
A	$$\dfrac{(R_1 C_1 s + 1)(R_2 C_2 s + 1)}{(R_1 C_1 s + 1)(R_2 C_2 s + 1) + R_2 C_1 s}$$	
B	$$\dfrac{R_1 + \dfrac{1}{C_1 s}}{\dfrac{1}{\dfrac{1}{R_2} + C_2 s} + R_1 + \dfrac{1}{C_1 s}}$$	
C	Both the answers are correct	
D	Both the answers are wrong	
53	If two second order systems have same damping ratio. What can you say about these two systems?	
A	They have the same maximum overshoot.	
B	Their settling times are equal	
C	Their natural frequencies must be the same.	
D	Their damped natural frequency must be the same.	
54	Figure shows the unit-step-response of a system. What can you say about the system? 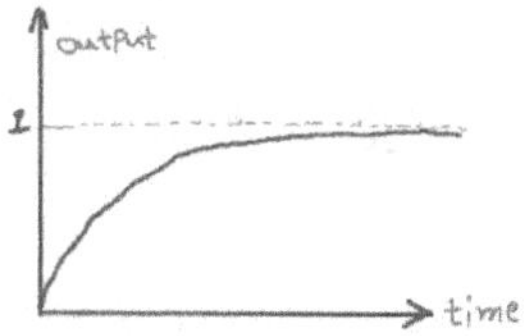	
A	Because there are no oscillations, it must be a first order system.	
B	Because there are no oscillations, it must be an underdamped second order system.	
C	Because there are no oscillations, it must be a second order system with very very large damping.	
D	None of the above it true.	
55	Mercury in glass thermometers used for measuring body temperature fall into the category of:	
A	First order systems with time constant in the range of 10-50 seconds.	

B	First order systems withe time constant in the range of 1-2 hours.
C	Second order systems with very small natural frequency and very small damping ratio.
D	Second order systems with large natural frequency and small damping ratio.
56	Figure shows the response of an unknown electrical circuit when subjected to 10V step input. An estimate of the transfer function of the circuit is:

A	$$\dfrac{3.74}{s^2 + 2.28s + 3.74}$$
B	$$\dfrac{10}{s^2 + 2.54s + 10}$$
C	$$\dfrac{s^2 + 1.24s + 2.36}{2.1s^2 + 1.73s + 2.74}$$
D	$$\dfrac{2.61s^2 + 1.32s + 2.45}{2.56s^2 + 1.73s + 2.87}$$
57	Pick the value of K_b so that the damping ratio of the closed loop system shown in Figure is 0.5.

A	0.216
B	0.657
C	0.234
D	1.236
58	A closed loop system with input as a voltage and output as a voltage has a transfer function of $\dfrac{36}{s^2+2s+36}$. The 2% settling time of this system when the input is a step of magnitude 10V is:

A	4 seconds
B	20 seconds
C	10 seconds
D	0.1 seconds
59	A closed loop system with input as a voltage and output as a voltage has a transfer function of $\frac{36}{s^2+2s+36}$. The rise time of this system when the input is a step of magnitude 10V is:
A	≈0.3 seconds
B	≈2 seconds
C	≈10 seconds
D	≈0.001 seconds
60	A closed loop system with input as a voltage and output as a voltage has a transfer function of $\frac{36}{s^2+2s+36}$. The % Peak overshoot of this system when the input is a step of magnitude 10V is:
A	$\approx59\%$
B	$\approx79\%$
C	$\approx29\%$
D	$\approx19\%$
61	A first-order, type-0 system is subjected to a unit-step input. The steady-state error is
A	Zero
B	$\frac{1}{1+K}$
C	Infinity
D	None of the above.
62	A first-order, type-1 system is subjected to a unit-step input. The steady-state error is
A	Zero
B	$\frac{1}{1+K}$
C	Infinity
D	None of the above.
63	A first-order, type-0 system is subjected to a unit-ramp input. The steady-state error is

A	Zero
B	$\dfrac{1}{1+K}$
C	Infinity
D	None of the above.
64	A first-order, type-1 system is subjected to a unit-ramp input. The steady-state error is
A	$\dfrac{1}{K}$
B	$\dfrac{1}{1+K}$
C	Infinity
D	None of the above.
65	A first-order, type-1 system is subjected to an input and the steady state error is seen to be very very large. The input could be:
A	A Sinusidal input
B	A step input
C	A ramp input
D	A parabolic input
66	A second-order, type-2 system is subjected to a parabolic input. The steady state error is:
A	$\dfrac{1}{K}$
B	$\dfrac{1}{1+K}$
C	Infinity
D	None of the above.
66	The equation of motion for a vibrating system with viscous damping is $$\dfrac{d^2x}{dt^2} + \dfrac{b}{m}\dfrac{dx}{dt} + \dfrac{k}{m}x = 0$$ If the roots of the characteristic equation are real, then the system will be
A	An overdamped system.
B	An underdamped system.
C	A critically damped system.
D	None of the above.
68	Which statement is most inaccurate comment about velocity error?

A	It is steady state error for a ramp input.
B	The dimension of the velocity error is same as the system error.
C	Velocity error is an error in velocity.
D	It is tracking error while tracking a ramp input.
69	A system has an Open Loop Transfer Function $G(s) = \frac{(1+\tau_1)(1+\tau_2)\cdots(1+\tau_m)}{s^3(1+T_1)(1+T_2)\cdots(1+T_n)}.$ The type of the system is:
A	Type 3 system
B	Type 2 system
C	Type m system
D	Type n system
70	A system has an Open Loop Transfer Function $$G(s) = \frac{(1 + \tau_1)(1 + \tau_2) \cdots (1 + \tau_m)}{s^3(1 + T_1)(1 + T_2) \cdots (1 + T_n)}$$ where τ1, τ, …,τm, T1, T2, …, Tn are real constants. The order and type of the system are:
A	Type = 3 and order = 3
B	Type = 3 and order = 3+n
C	Type = 3 and order = 3+m
D	Type = 3 and order = 2
71	A system has an Open Loop Transfer Function $$G(s) = \frac{(1 + \tau_1 s)(1 + \tau_2 s) \cdots (1 + \tau_m s)}{s^3(1 + T_1 s)(1 + T_2 s) \cdots (1 + T_n s)}$$ where $\tau_1, \tau_2, \ldots, \tau_m, T_1, T_2, \ldots, T_n$ are real constants. The order and type of the system are:
A	Type = 3 and order = 3
B	Type = 3 and order = 3+n
C	Type = 3 and order = 3+m
D	Type = 3 and order = 2
72	A unity feedback closed loop system has a Transfer Function $$\frac{(1 + \tau_1 s)(1 + \tau_2 s) \cdots (1 + \tau_m s)}{(1 + \tau_1 s)(1 + \tau_2 s) \cdots (1 + \tau_m s) + s^2(1 + T_1 s)(1 + T_2 s) \cdots (1 + T_n s)}$$ where $\tau_1, \tau_2, \ldots, \tau_m, T_1, T_2, \ldots, T_n$ are real constants. The order and type of the system are:
A	Type = 2 and order = 2
B	Type = 3 and order = 2+n

C	Type = 2 and order = 2+n
D	Type = 3 and order = 2
73	The static position error of a system with open loop transfer function of G(s) is:
A	$\lim_{s \to 0} G(s)$
B	$\lim_{s \to 0} sG(s)$
C	$\lim_{s \to 0} s^2 G(s)$
D	None of the above.
74	The static Velocity error of a system with open loop transfer function of G(s)is:
A	$\lim_{s \to 0} sG(s)$
B	$\lim_{s \to 0} G(s)$
C	$\lim_{s \to 0} s^2 G(s)$
D	None of the above.
75	The static Acceleration error of a system with open loop transfer function of G(s)is:
A	$\lim_{s \to 0} s^2 G(s)$
B	$\lim_{s \to 0} G(s)$
C	$\lim_{s \to 0} sG(s)$
D	None of the above.
76	Which statement given below is most inaccurate one about ideal operational amplifiers?
A	An ideal op amp has zero input impedance.
B	An ideal op amp has zero output impedance.
C	An ideal op amp does not draw any current.
D	An ideal op amp can work with both ac and dc voltages.
77	Which statement given below is most appropriate one about ideal operational amplifiers?
A	An ideal op amp has infinite input impedance.
B	An ideal op amp has infinite output impedance
C	An ideal op amp draws lot of current into its terminals.
D	An ideal op amp cannot support currents in load.
78	The equation of motion for a vibrating system with viscous damping is $$\frac{d^2x}{dt^2} + \frac{b}{m}\frac{dx}{dt} + \frac{k}{m}x = 0$$ If the roots of the characteristic equation are real and equal, then the system will be

A	An overdamped system.
B	An underdamped system.
C	A critically damped system.
D	None of the above.
79	Which statement is most inaccurate one about an actual op amp?
A	Actual op amps draw less current than an ideal one.
B	Actual op amps draw a bit of current.
C	Actual op amp is always inverts the input.
D	Actual op amp never negates the input.
80	The transfer function of the following op amp circuit is:
A	$-\dfrac{R_2}{R_1} s$
B	$-\dfrac{R_1}{R_2} s$
C	$-\dfrac{R_1}{R_2}$
D	$-\dfrac{R_2}{R_1}$
81	The transfer function of the following op amp circuit is:
A	$\left(\dfrac{R_1 + R_2}{R_1}\right)$
B	$\left(\dfrac{R_1 + R_2}{R_2}\right)$
C	$\left(\dfrac{R_1}{R_1 + R_2}\right)$
D	$\left(\dfrac{R_2}{R_1 + R_2}\right)$

82	The transfer function $\dfrac{E_o(s)}{E_i(s)}$ for the op amp circuit shown in figure is:
A	$-\dfrac{Z_2(s)}{Z_1(s)}$
B	$-\dfrac{Z_1(s)}{Z_2(s)}$
C	$-\dfrac{Z_1(s) + Z_2(s)}{Z_2(s)}$
D	$-\dfrac{Z_1(s) + Z_2(s)}{Z_1(s)}$
83	The transfer function of the cascaded op amp circuit is:
A	$-\dfrac{R_2 R_4}{R_1 R_3}\dfrac{R_1 C_1 s + 1}{R_2 C_2 s + 1}$
B	$\dfrac{R_2 R_4}{R_1 R_3}\dfrac{R_1 C_1 s + 1}{R_2 C_2 s + 1}$
C	$\dfrac{R_2 R_4}{R_1 R_3}\dfrac{R_1 C_1 s}{R_2 C_2 s + 1}$
D	None of the above
84	Which of the following statements aboout PD control is inaccurate one?
A	PD controller adds a zero at $-\dfrac{K_P}{K_D}$.
B	PD controller in general increases damping.
C	PD controller decreases maximum overshoot.
D	PD controller removes constant steady-state error.
85	Which of the following statements about PI control is an accurate representation?
A	When PI controller is used, the system type is increased by 1.
B	PI controller cannot remove constant steady state error.
C	Any choice of Kp and Ki of PI controller will stabilise the system.
D	PI control in general reduces the damping in system.

87	The equation of motion for a vibrating system with viscous damping is

$$\frac{d^2x}{dt^2} + \frac{b}{m}\frac{dx}{dt} + \frac{k}{m}x = 0$$

If the roots of the characteristic equation are complex conjugate, then the system will be

A	An overdamped system.
B	An underdamped system.
C	A critically damped system.
D	None of the above.

88	Consider following statements about an ideal op amp:

   (i)   It has infinite input impedance.

   (ii)   It has zero output impedance.

   (iii)   Any load can be connected to it.

   (iv)   It draws lot of input current.

   (v)   It doesn't draw lot of input current.

   (vi)   It has infinite output impedance.

Which of the following choices is most appropriate?

A	Only statements (i) and (ii) are true.
B	Only statement (ii) is false.
C	Only statement (iv) is false.
D	Statements (i), (ii), (iii), (v) are true and others are false.
89	Which statement is most inaccurate comment about velocity error?
A	It is steady state error for a ramp input.
B	The dimension of the velocity error is same as the system error.
C	Velocity error is an error in velocity.
D	It is tracking error while tracking a ramp input.

90	The equation of motion for a vibrating system with viscous damping is

$$m\frac{d^2x}{dt^2} + b\frac{dx}{dt} + kx = 0$$

Damping ratio for this system is:

A	$\dfrac{2b}{\sqrt{mk}}$
B	$\dfrac{b}{2\sqrt{mk}}$
C	$\dfrac{b}{\sqrt{mk}}$
D	None of the above.

91	The equation of motion for a vibrating system with viscous damping is $$m\frac{d^2x}{dt^2} + b\frac{dx}{dt} + kx = 0$$ Damping ratio for this system is:
A	$\dfrac{2b}{\sqrt{mk}}$
B	$\dfrac{b}{2\sqrt{mk}}$
C	$\dfrac{b}{\sqrt{mk}}$
D	None of the above.
92	With zero volts on both inputs, an OP-amp ideally should have an output
A	equal to positive supply voltage.
B	equal to the negative supply voltage
C	equal to zero
D	Equal to CMRR
93	Of the values listed, the most realistic value for open-loop voltage gain of an OP-amp is ......
A	1
B	200
C	500
D	1,00,000
94	If the values of resistances are $R_1 = 1\Omega$, $R_2 = 100\Omega$, then the overall gain of the op amp shown in figure is:
A	1,00,001
B	100
C	101
D	-101
95	A certain inverting amplifier has a closed-loop voltage gain of 25. The Op-amp has an open-loop voltage gain of 100,000. If an Op-amp with an open-loop voltage gain of 200,000 is substituted in the arrangement, the closed-loop gain is:

A	25
B	12.5
C	50
D	25.5
96	A certain inverting amplifier has a closed-loop voltage gain of 25. The Op-amp has an open-loop voltage gain of 100,000. If an Op-amp with an open-loop voltage gain of 200,000 is substituted in the arrangement, the closed-loop gain is:
A	25
B	12.5
C	50
D	25.5
96	Consider the assertions about the block diagram. Which of them represents the precise condition? (A). Block diagram is used for analysis and design of control systems. (B). Block diagram also provides information regarding the physical construction of the system.
A	(B) is true but (A) is false.
B	(A) is true but (B) is false.
C	Both (A) and (B) are true.
D	Both (A) and (B) are false.
97	If finite number of blocks are connected in series or cascade configuration in a block diagram, then how are the blocks combined algebraically?
A	By addition.
B	By subtraction.
C	By division.
D	By multiplication.
98	Consider the block diagram in figure. The transfer function of this block diagram is:

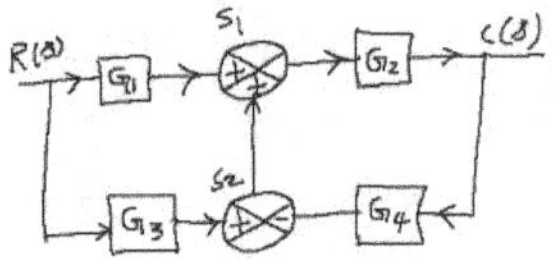

A	$$\dfrac{G_2 G_4 G_3}{1 + G_2 G_1}$$
B	$$\dfrac{G_2 (G_4 + G_3)}{1 + G_2 G_1}$$
C	$$\dfrac{G_2 (G_1 + G_3)}{1 + G_2 G_4}$$
D	$$\dfrac{G_2 G_4 G_3}{1 + G_2 G_1 G_4}$$
99	Consider the block diagram in figure. Which block diagram is equivalent to this block diagram?
A	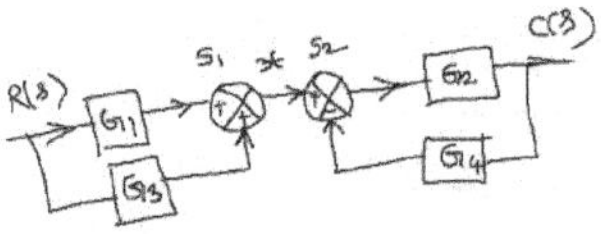
B	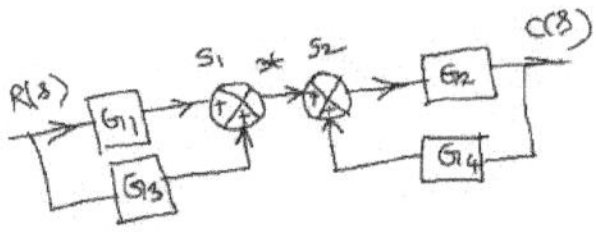
C	

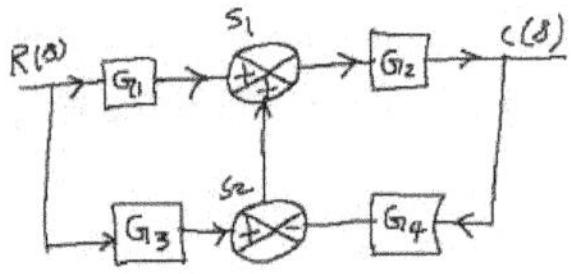

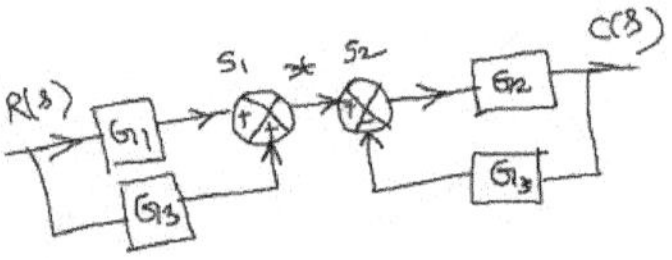

<table>
<tr><td>D</td><td>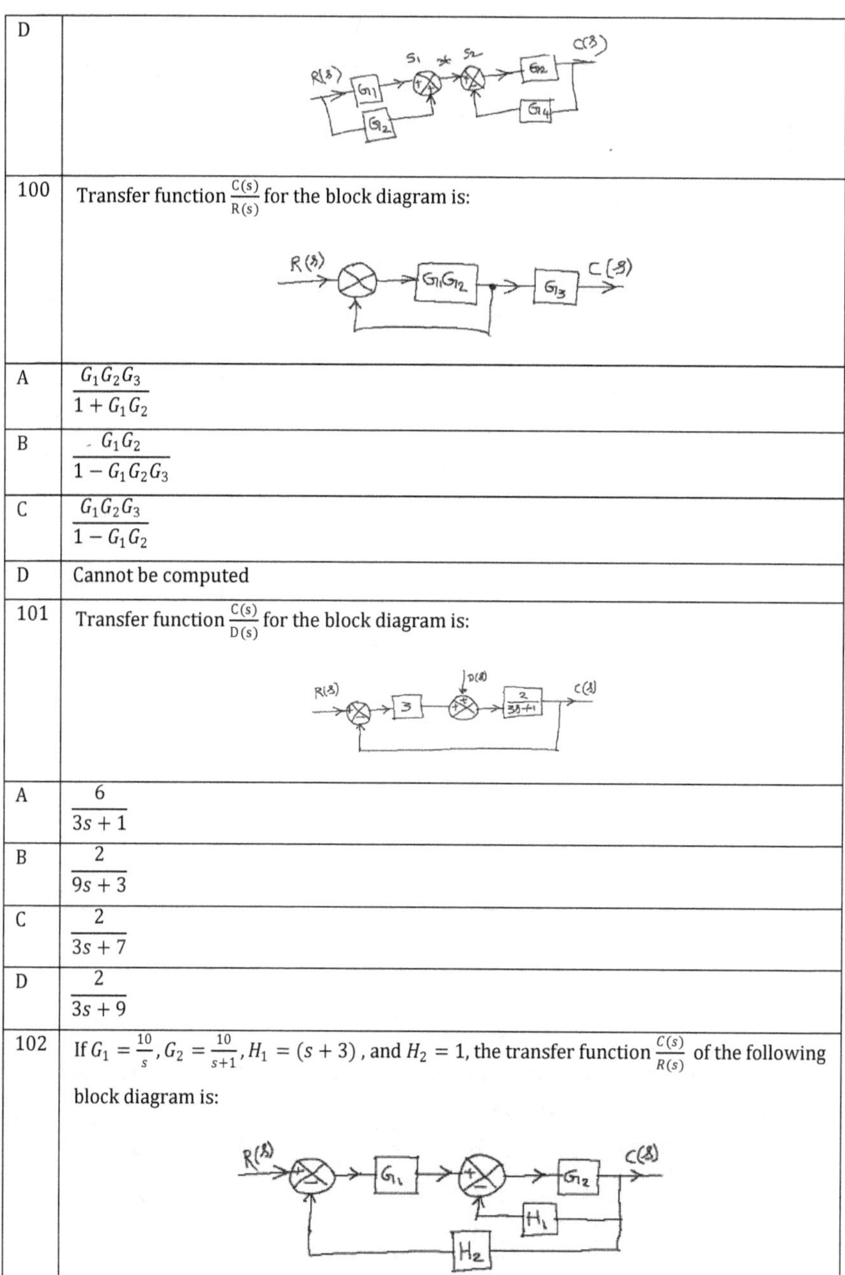
</td></tr>
</table>

100	Transfer function $\frac{C(s)}{R(s)}$ for the block diagram is:

A	$\dfrac{G_1 G_2 G_3}{1 + G_1 G_2}$
B	$\dfrac{G_1 G_2}{1 - G_1 G_2 G_3}$
C	$\dfrac{G_1 G_2 G_3}{1 - G_1 G_2}$
D	Cannot be computed

101	Transfer function $\frac{C(s)}{D(s)}$ for the block diagram is:

A	$\dfrac{6}{3s + 1}$
B	$\dfrac{2}{9s + 3}$
C	$\dfrac{2}{3s + 7}$
D	$\dfrac{2}{3s + 9}$

102	If $G_1 = \dfrac{10}{s}, G_2 = \dfrac{10}{s+1}, H_1 = (s + 3)$, and $H_2 = 1$, the transfer function $\dfrac{C(s)}{R(s)}$ of the following block diagram is:

A	$\dfrac{10}{11s^2 + 31s + 10}$
B	$\dfrac{100}{11s^2 + 31s + 100}$
C	$\dfrac{100}{10s^2 + 31s + 100}$
D	$\dfrac{10}{10s^2 + 31s + 100}$
103	When deriving the transfer function of a linear element:
A	Both initial conditions and loading are taken into account.
B	Initial conditions are taken into account but the element is assumed to be not loaded.
C	Initial conditions are assumed to be zero but loading is taken into account
D	Initial conditions are assumed to be zero and the element is assumed to be not loaded
104	The transfer function of the block diagram shown below is:
A	$\dfrac{G_1 + G_2}{1 - G_1 H_1}$
B	$\dfrac{G_1 + G_2}{1 + G_1 H_1}$
C	$\dfrac{G_1 + G_2}{1 - G_2 H_1}$
D	$\dfrac{G_1 + G_2}{1 + G_1 H_1}$
105	 In an attempt to find the transfer function, the branch point B1 is moved to a position after the summing point S1 as shown by the dotted lines. Of the four choices given pick the correct block diagram obtained after this reduction step.

A	R S_1 B_1 G_1 S_2 B_2 G_2 y S_{new} H_1
B	R S_1 B_1 G_1 S_2 B_2 G_2 y S_{new} H_1

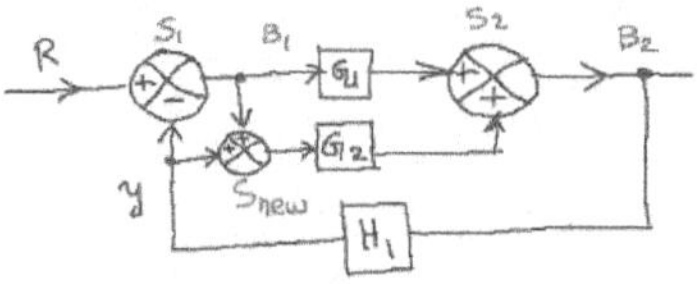

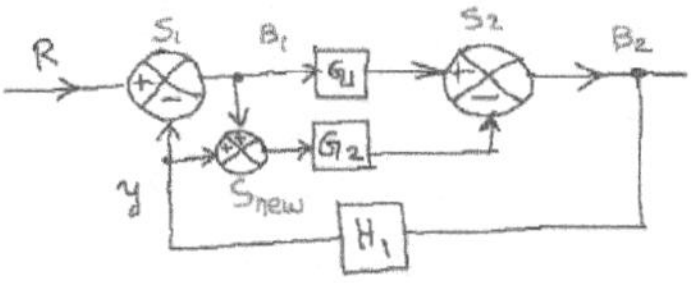

C	If we remove the summing point S_{new} from option A, it will be correct answer.
D	If we remove the summing point S_{new} from option B, it will be correct answer.
106	Which system is also known as an automatic control system?
A	Open loop control system
B	Closed loop Control system
C	A system for which the mathematical block diagram has a loop in int.
D	None of the above.
107	Which of the following is the electrical analogous quantity for mass?
A	Current
B	Resistance
C	Capacitance
D	Inductance
108	Which of the following is mechanical analogous quantity for current?
A	Spring compliance
B	Spring stiffness
C	Velocity
D	Acceleration
109	Which of the following are the characteristics of negative feedback systems?
A	Low sensitivity to parameter variations.
B	Rejection of disturbance signals.
C	Improve ment in bandwidth.
D	All above.
110	Which of the following statement is most appropriate?
A	Type of system is obtained from Open Loop Transfer Function.

B	Steady state analysis depends on type of the system.
C	Transient analysis depends on the order of the system.
D	All above are true.
111	A system is defined by the following dynamic equation: $4\dfrac{d^2c(t)}{dt^2} + 8\dfrac{dc(t)}{dt} + 16c(t) = 16u(t)$ where u(t) is the input to the system. The order of the system is:
A	2
B	3
C	1
D	None of the above.
112	A linear time invariant system, when subject to a unit step input, $u_s(t)$ gave a response of $c(t) = te^{-t}$. The transfer function of the system is:
A	$\dfrac{s}{(s+1)}$
B	$\dfrac{s}{(s+1)^2}$
C	$\dfrac{1}{s(s+1)^2}$
D	$\dfrac{1}{(s+1)^2}$
113	Who proposed the gain formula for signal flow graphs?
A	S.J. Mason
B	Benjamin C. Kuo
C	Katsushito Ogata
D	R.E. Evans
114	In which year was gain formula for signal flow graphs was presented?
A	1990's
B	1970's
C	1950's
D	1940's
115	Consider the following assertions regarding a node in a signal flow graph: (A). Every node acts as a summing point. (B). There will be as many nodes as the number of variables in system's equations. (C). Mason's gain formula is applicable to all signal flow graphs.

	(D). Every signal flow graph will compulsorily have an input node and an output node or we can easily create one. Pick the most appropriate option from the choices given.
A	Only (A) and (B) are correct.
B	Only (B) and (C) are correct.
C	Only (C) and (D) are correct.
D	(A), (B), (C), and (D) are all correct.
116	The determinant of a signal flow graph depends on:
A	Only the number of loops.
B	The number, topology, and interrelations of loops.
C	The number of forward paths
D	Loops, topology, interrelations and forward paths.
117	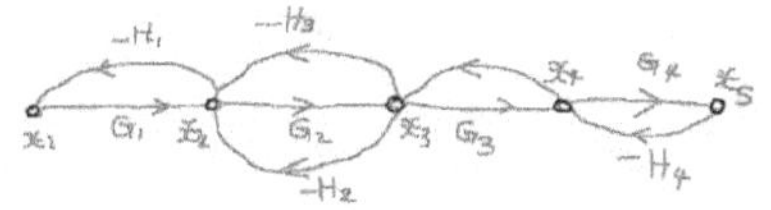 Consider the signal flow graph given in figure. Which choice is equivalent to the given graph?
A	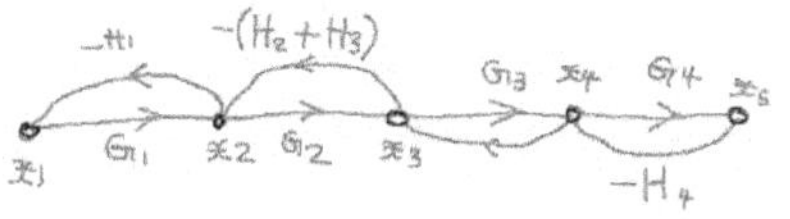
B	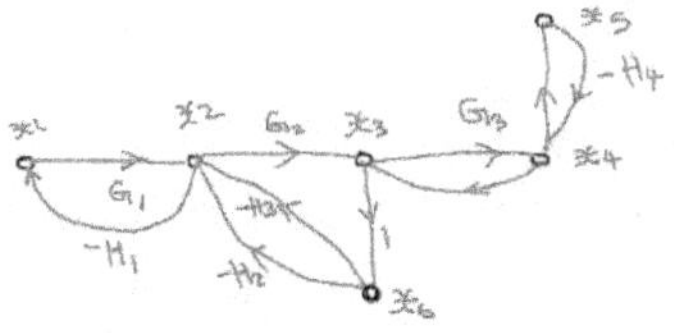
C	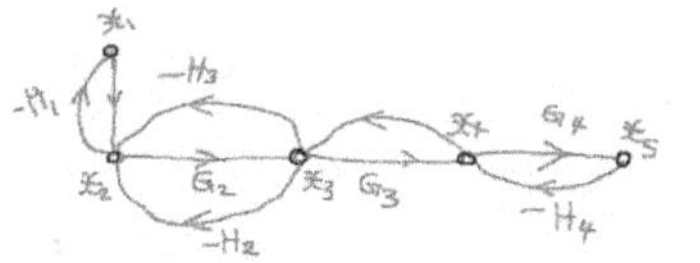
D	All the above

| 118 | Cosndier the signal flow graph given in the figure. Pick the most appropriate comment about the sfg. |

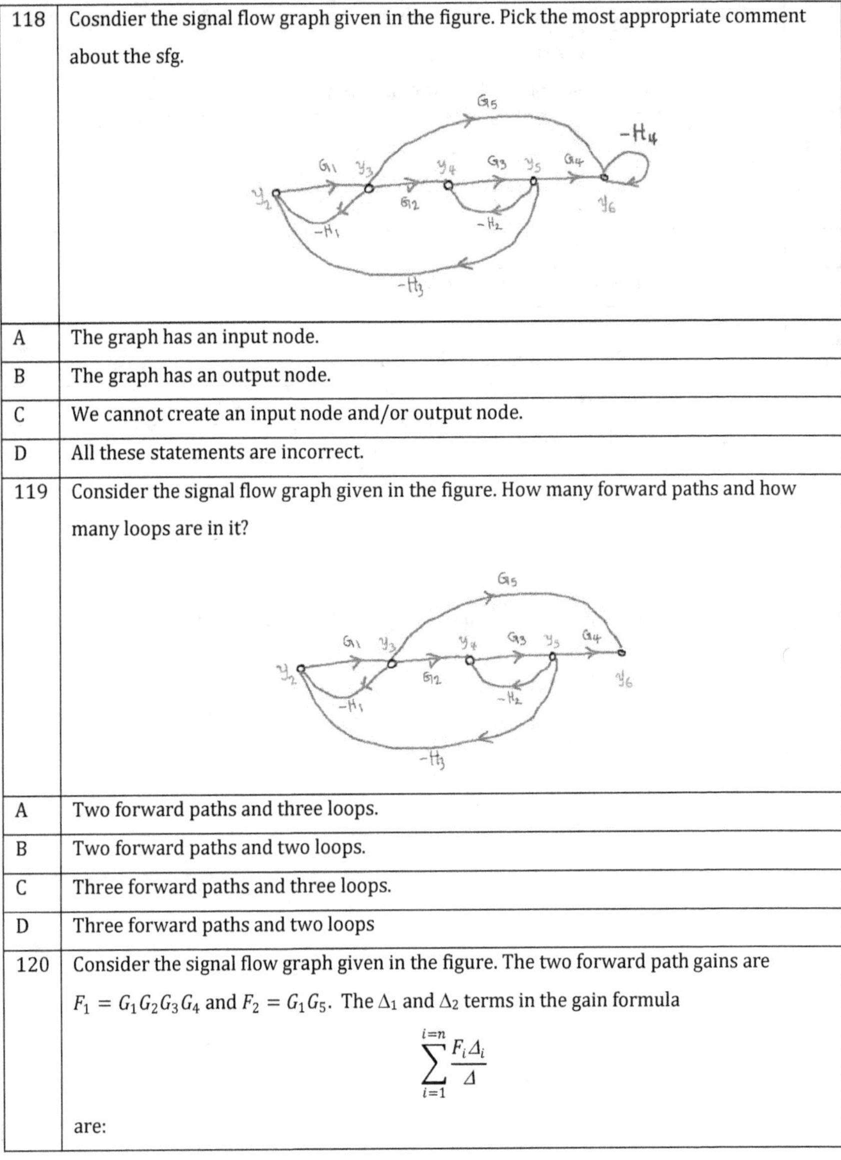

A	The graph has an input node.
B	The graph has an output node.
C	We cannot create an input node and/or output node.
D	All these statements are incorrect.

| 119 | Consider the signal flow graph given in the figure. How many forward paths and how many loops are in it? |

A	Two forward paths and three loops.
B	Two forward paths and two loops.
C	Three forward paths and three loops.
D	Three forward paths and two loops

| 120 | Consider the signal flow graph given in the figure. The two forward path gains are $F_1 = G_1 G_2 G_3 G_4$ and $F_2 = G_1 G_5$. The Δ_1 and Δ_2 terms in the gain formula $$\sum_{i=1}^{i=n} \frac{F_i \Delta_i}{\Delta}$$ are: |

A	$\Delta_1 = 1$ and $\Delta_2 = 1 + G_3 H_2$
B	$\Delta_2 = 1$ and $\Delta_1 = 1 + G_3 H_2$
C	$\Delta_1 = 1 + G_1 H_1$ and $\Delta_2 = 1 + G_3 H_2$
D	$\Delta_2 = 1 + G_1 H_1$ and $\Delta_1 = 1 + G_3 H_2$
121	A traffic signal operated on a time basis is an example for ________
A	Open loop Control System
B	Closed loop Control System
C	Negative feedback Control System
D	Positive feedback Control System
122	The transfer function of the following block diagram is ________
A	GH
B	G/1+GH
C	G/1-GH
D	G-H
123	The Laplace transform of d^2x/dt^2 is ______
A	$s^2 X(s)$
B	$s X(s)$
C	$X(s)$
D	$X(s) / s^2$
124	The Laplace transform of $e^{-at} . f(t)$ is ________
A	F(s-a)

B	F(s)
C	F(a)
D	F(s+a)
125	When a human being tries to approach an object, his brain acts as ________
A	An error measuring device
B	A controller
C	An actuator
D	An amplifier
126	Consider the following statements: 1. The effect of feedback is to reduce the system error 2. Feedback increases the gain of the system in one frequency range but decreases in another 3. Feedback can cause a system that is originally stable to become unstable Which of these statements are correct?
A	1, 2 and 3
B	1 and 2
C	2 and 3
D	1 and 3
127	A negative-feedback closed-loop system is supplied to an input of 5V. The system has a forward gain of 1 and a feedback gain of a 1. What is the output voltage?
A	1.0V
B	1.5V
C	2.0V
D	2.5V
128	The transfer function of a linear-time-invariant system is given as $1/(s+1)$. What is the steady-state value of the unit-impulse response?
A	0
B	1
C	2
D	Infinite
129	The systems in which the output has no effect on the control action are called as ________ control systems.
A	Open loop

B	Closed loop
C	Feedback loop
D	None of the above
130	In a closed loop control system with positive value of feedback gain, the overall gain of the system will _______
A	Decrease
B	Increase
C	Will not change
D	None of the above
131	In an open loop system,
A	The control action depends on the size of the system
B	The control action is independent of the output
C	The control action depends on the input signal
D	The control action depends on the system variables
132	A transfer function has two zeroes at infinity. Then the relation between the numerator(N) and the denominator degree(M) of the transfer function is:
A	$N=M+2$
B	$N=M-2$
C	$N=M+1$
D	$N=M-1$
133	A synchro Transmitter is used with control transformer for:
A	Feedback
B	Error detection
C	Amplification
D	Remote Sensing
134	The output of the feedback control system must be a function of:
A	Reference input
B	Reference output
C	Output and feedback signal
D	Input and feedback signal
135	The overall transfer function from block diagram reduction for cascaded blocks is _______
A	Sum of Individual gain
B	Difference of Individual gain

C	Product of Individual gain
D	Division of Individual gain
136	The overall transfer function of two blocks in parallel are :
A	Sum of Individual gain
B	Difference of Individual gain
C	Product of Individual gain
D	Division of Individual gain
137	A node having only outgoing branches.
A	Input Node
B	Output Node
C	Sink Node
D	Summing point
138	A node having only incoming branches.
A	Input Node
B	Output Node
C	Source Node
D	Summing point
139	The closed system has higher _______ than open loop control system, this implies increased speed of response.
A	Gain
B	Bandwidth
C	Frequency
D	Speed
140	_________ signal becomes zero, when the feedback signal and reference signals are equal.
A	Input
B	Actuating
C	Feedback
D	Reference
141	*Force balance equation of a Mass element is _______ (Assume displacement is 'x')*
A	M dx/dt
B	M^2 dx/dt
C	M d^2x/dt^2
D	M^2 d^2x/dt^2

142	Basically, poles of transfer function are the Laplace transform variable values which causes the transfer function to become _________
A	Zero
B	Unity
C	Infinite
D	Average Value
143	What is the value of parabolic input in Laplace domain?
A	1
B	A/S^3
C	A/S^2
D	A/S
144	For a first order system having transfer function 1/(1+sT), the unit impulse response is _________
A	$e^{-t/\tau}$
B	$\dfrac{1}{\tau}e^{-\left(\frac{t}{\tau}\right)}$
C	$\tau.e^{-t/\tau}$
D	$e^{-t/2\tau}$
145	In a feedback control system, the controller gets its input from __________
A	Load variable
B	Manipulated variable
C	Controlled Variable
D	Inferred Variable
146	For an undamped system, the value of the damping coefficient is _______
A	1
B	<1
C	0
D	>1
147	For a critically damped system, the value of the damping coefficient is _______
A	1
B	<1
C	0
D	>1

148	The inverse Laplace transform of the function f(s) is 1/s(s+1) is _______
A	$1-e^{-t}$
B	$1+e^{-t}$
C	e^{-t}
D	e^{t}
149	Force balance equation for elastic element (K) is (where x = displacement)
A	$K\,d^2x/dt^2$
B	$K.\,dx/dt$
C	$K.x$
D	$K.x^2$
150	The time taken for the response to reach half of the final value, for the very first time is called as ________ .
A	Rise time
B	Fall time
C	Delay time
D	Settling time
151	The time taken for the response to reach and stay within a specified tolerable error rate is called as _________
A	Rise time
B	Fall time
C	Delay time
D	Settling time
152	The time taken for the response to raise from 10% to 90% of the final value is called as _______
A	Rise time
B	Fall time
C	Delay time
D	Settling time
153	If the system response is completely oscillatory in nature throughout the entire time period, then the system is said to _________ system.
A	Undamped
B	Underdamped
C	Critically damped
D	Over damped

154	The _________ response of a system changes, when the input changes from one state to another.
A	Steady state
B	Transient
C	Steady state error
D	None
155	In a _________ signal, the instantaneous value varies as square of time from an initial value of zero at t=0.
A	Step
B	Ramp
C	Parabolic
D	Impulse
156	The _________ response of a system can be called as a weighting function.
A	Step
B	Ramp
C	Parabolic
D	Impulse
157	The _________ is the value of the error signal e(t), when t tends to infinity.
A	Steady state error
B	Infinite state error
C	Finite state error
D	Type error
158	On which factor does the steady state error of the system depend?
A	Order
B	Type
C	Size
D	Prototype
159	In time domain system, which response has its existence even after an extinction of transient response?
A	Step
B	Impulse
C	Steady State
D	None

160	The step error coefficient of a system $G(s) = 1 \ (S+6)(S+1)$ with unity feedback is
A	1/6
B	0
C	1
D	∞
161	The poles of a continuous time oscillator are ______
A	Real
B	Imaginary
C	Complex conjugate
D	None
162	For a second order system, damping ratio $(\xi), is\ 0 < \xi < 1$, then the roots of the characteristic polynomial are ______
A	Complex Conjugates
B	Real and equal
C	Real but not equal
D	Imaginary
163	If $L[f(t)] = 2(S+1)/(\ S^2+2S+5)\ then\ f(0\ +)\ and\ f(\infty)$ are given by
A	0 , 2 respectively
B	2 , 0 respectively
C	2/5, 0 respectively
D	0 , 1 respectively
164	The final value theorem is used to find the ______
A	Initial value of the system
B	Rise time of the system
C	Transient behavior of the system
D	Steady state value of the system
165	Gain of the signal flow graph shown in the figure is: 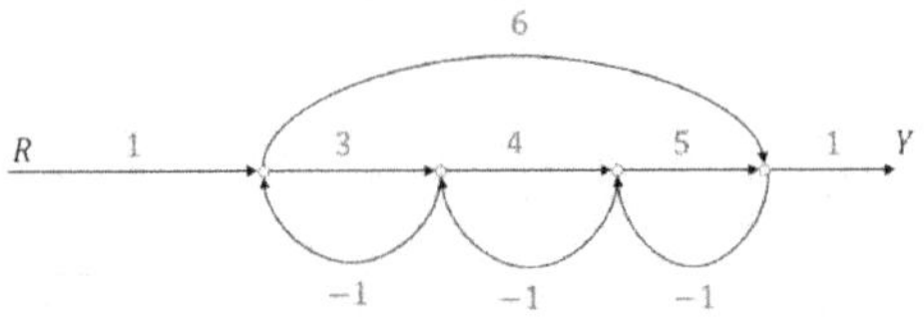

A	$$\dfrac{45}{17}$$
B	$$\dfrac{44}{23}$$
C	$$\dfrac{66}{19}$$
D	$$\dfrac{90}{19}$$

166	Transfer function of the system with the following signal flow graph is: 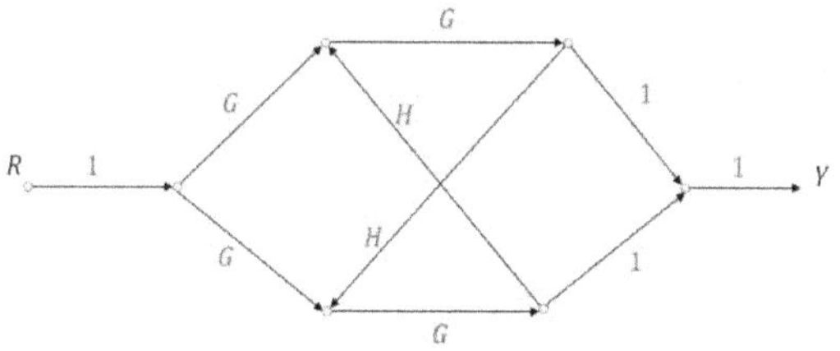

A	$$\dfrac{2G^2 + 2G^3 H^1}{1 + G^2 H^2}$$
B	$$\dfrac{2G^2 + 2G^3 H^1}{1 - G^2 H^2}$$
C	$$\dfrac{2G^3 + 2G^3 H^1}{1 - G^2 H^2}$$
D	$$\dfrac{2G^2 + 2G^3 H^2}{1 - G^2 H^2}$$

167	Transfer function of the system with the block diagram shown below is: 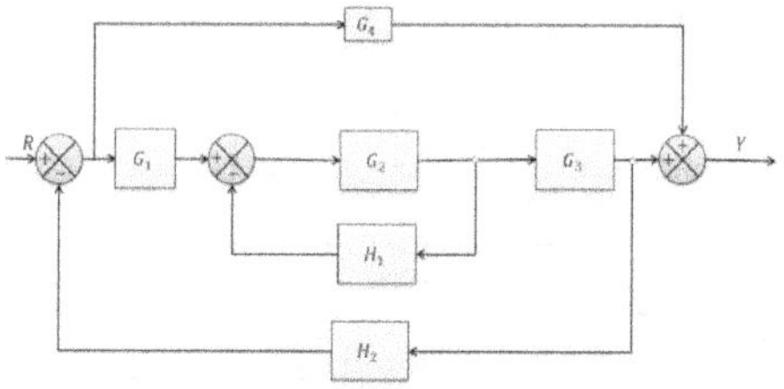

A	$$\dfrac{G_1 G_2 G_3 + G_4(1 + G_2 H_1)}{1 + G_2 H_1 + G_1 G_2 G_3 H_2 + G_4 H_1}$$
B	$$\dfrac{G_1 G_2 G_3 H_1 + G_4(1 + G_2 H_1)}{1 + G_2 H_1 + G_1 G_2 G_3 H_2}$$
C	$$\dfrac{G_1 G_2 G_3 + G_4(1 + G_2 H_1)}{1 + G_2 H_1 + G_1 G_2 G_3 H_2}$$

168	What is the impulse response of a system with the transfer function $\dfrac{1}{(s+1)(+3)^2}$?
A	$\dfrac{1}{2}e^{-t} - \dfrac{1}{4}e^{-3t} - \dfrac{1}{2}te^{-3t}$
B	$\dfrac{1}{4}e^{-t} - \dfrac{1}{4}e^{-3t} - \dfrac{1}{2}te^{-3t}$
C	$\dfrac{1}{2}e^{-t} + \dfrac{1}{4}e^{-3t} - \dfrac{1}{2}te^{-3t}$
D	$\dfrac{1}{2}e^{-t} - \dfrac{1}{4}e^{-3t} + \dfrac{1}{2}te^{-3t}$
169	Which of the following statement is true for the step response of a stable system, given t_d is the delay time, t_r is the rise time and t_s is the settling time ?
A	$t_s \leq t_r \leq t_d$
B	$t_s < t_r < t_d$
C	$t_r \leq t_s \leq t_d$
D	$t_d < t_r < t_s$
170	Match the following (1,2,3,4 to a,b,c,d): 1) Type of the system : 1 2) Type of the system : less than 2 3) Type of the system : 2 4) Type of the system : greater than 0 a) $K_a=0$ b) $K_p=\infty$ c) $K_v=$ constant (non-zero) d) $K_a=$ constant (non-zero)
A	1-c; 2-b; 3-a; 4-d
B	1-a; 2-d; 3-b; 4-c
C	1-c; 2-a; 3-d; 4-b
D	1-b; 2-a; 3-c; 4-d
171	A step input applied to a unity feedback system results in a steady state error of 0.05. What is the type of the system?
A	3
B	2
C	1
D	0

172	A step input applied to a unity feedback system results in a steady state error of **0.05**. What is the position error constant for this system?
A	20
B	19
C	18
D	17
173	Given an open loop transfer function $G(s) = \frac{25}{s^2+Ks}G(s)$. What is **K** given that the closed loop (unity feedback) response of the system is critically damped (unit step input)?
A	25
B	15
C	10
D	5
174	What is the value of K for a unity feedback (negative) system with open loop transfer function $G(s) = \frac{K}{s(s+1)(s+3)}$ to get e_{ss}=0.15 when a unit ramp input is applied?
A	20
B	30
C	40
D	15
175	What is the steady state error for a unity feedback (negative) system with open loop transfer function $G(s) = \frac{40}{s(s+2)(s+4)}$ when a unit step input is applied ?
A	$\dfrac{1}{5}$
B	$\dfrac{1}{4}$
C	$\dfrac{1}{6}$
D	5
176	What is the percentage of peak overshoot in the response of a system with the transfer function $\frac{Y(s)}{R(s)} = \frac{64}{s^2+64}$?
A	0
B	50
C	100
D	26

177	Can Routh's Array be used to comment on relative stability?
A	Yes. Routh-Hurwitz criterion is very powerful tool for analyzing relative stability.
B	Routh's criterion talks about only absolute stability.
C	Routh's criterion can give quantitative information on relative stability.
D	None of the above
178	Consider the assertions below: (i) Every system must be very very stable if it needs to be useful. (ii) If a system is more stable, then it is more useful. (iii) If a system is barely stable, it is sufficient if it meets performance criteria. (iv) If system (A) has better relative stability than system (B),then system (A) must be chosen and put to use. Which of the following options most accurately describes the assertions?
A	Options (i), (ii), (iii) are very accurate and option (iv) is not correct.
B	Options (iv) and (iii) are correct and others are incorrect.
C	All assertions given above are very accurate.
D	There has to be a trade-off between stability and flexibility and all assertions are inaccurate up to an extent.
179	If we consider cargo ships and warships, which statement below describes relative stability of ships?
A	Cargo ships need to be relatively stable and war ships need to be flexible without too much stability.
B	Both cargo ships and warships need to be equally and very stable.
C	As long as ships are stable, we need not bother about anything else.
D	Flexibility and usability are very important and we may compromise on stability.
180	Which expression below puts BIBO-stability of system S, with input u(t) and output y(t), most accurately into mathematical notation.
A	$[S \text{ is BIBO stable}] \Rightarrow [\{|u(t)| \leq M, t \geq 0\} \Rightarrow \{\exists N \ni |y(t)| \leq N, t \leq 0\}]$
B	$[S \text{ is BIBO stable}] \Rightarrow [\exists M, N \in R \ni |u(t)| \leq M \wedge |y(t)| \leq N, t \geq 0]$
C	$[S \text{ is BIBO stable}] \Rightarrow [\forall M, \{|u(t)| \leq M, t \geq 0\} \Rightarrow \{\exists N \ni |y(t)| \leq N, t \geq 0\}]$
D	$[S \text{ is BIBO stable}] \Rightarrow [\{|u(t)| \leq M, t \geq 0\} \Rightarrow \{\exists N \ni |y(t)| \leq N, t \geq 0\}]$
181	Which statement below is most pertinent comment about Routh's stability criterion?
A	Routh's criterion merely specifies whether a system is stable or not.
B	Routh's criterion may be used to find the range of values of system gain for which the stability of the system is ensured.

C	Systems with pure time-delay cannot be handled by Routh's criterion for stability analysis without resorting to some type of approximation for the time-delay.
D	All the above statements are pertinent.
182	Pick the statement that is most appropriate:
A	Asymptotic stability is more stringent requirement than exponential stability.
B	Exponential stability is more stringent requirement than asymptotic stability which is more stringent requirement than stability in the sense of Lyapunov.
C	Global asymptotic stability is less stringent requirement than stability in the sense of Lyapunov
D	Uniform asymptotic stability in the sense of Lyapunov is the most stringent requirement.
183	What goes into the blank portion of the following mathematical statement regarding a dynamical system $\dot{x} = f(x,t)$. $B(x,r)$ represents a ball of radius r centered at x. $$[\underline{\quad}] \Leftrightarrow [\exists r > 0 \ni \nexists x \in B(x_e, r) \ni \{(f(x,t) = 0), \forall\, t \geq t_1 \wedge (x \neq x_e)\}]$$
A	The system is exponentially stable.
B	The system is asymptotically stable.
C	x_e is isolated equilibrium.
D	Stability and Asymptotic stability are equivalent.
184	If an equilibrium point is attractive then:
A	It is stable and bounded.
B	It is stable and solution tends to equilibrium point as t→∞.
C	The system has magnets and attracts everything else.
D	The equilibrium point tends to zero as t→∞.
185	Pick the most appropriate statement:
A	A fighter aircraft is not very stable since it has to be unpredictable dynamical system.
B	A fighter aircraft has to appear as erratic to the enemy hence fighter aircrafts are designed to be very unstable.
C	A fighter aircraft must be able to perform evasive maneuvers in hostile situations and hence it is designed to be relatively less stable than a commercial aircraft.
D	Pilots are thoroughly trained to fight tough battles hence stability concerns are NOT at all considered while designing fighter aircrafts.
186	For the system shown in Figure, what value of K will result in sustained oscillations?

	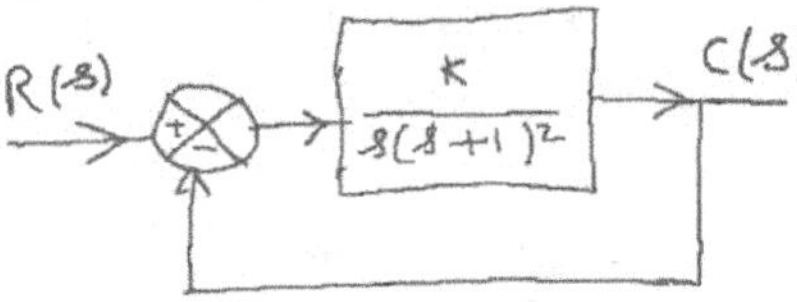 R(s) $\dfrac{K}{s(s+1)^2}$ C(s)
A	K=1
B	K=2
C	K=3
D	K=4
187	A feedback control system has open loop transfer function $G(s) = \dfrac{K}{s(s+1)(s+2)}$. Which of the following values of K is desirable to avoid even though all the roots of the characteristic equation are in the left half of the s-plane?
A	K=0.273
B	K=0.192
C	K=0.384
D	K=0.325
188	A feedback system has Open Loop Transfer Function $G(s) = \dfrac{K}{s(s+2)(s+1)}$. For K=0.384, the factored form of characteristic equation is:
A	(s+1.7)(s+2.3)(s+0.098)
B	(s+2.3)(s+0.7)(s+0.24)
C	(s+3.7)(s+0.98)(s+0.11)
D	(s+0.422)²(s+2.156)
189	A system has open loop transfer function $G(s) = \dfrac{K(s+5)}{s(s+1)(s+3)}$. For which value of K does the closed loop system with unity feedback have tendency to exhibit sustained oscillations at 4 rad/sec frequency?
A	K=0.
B	K>3.
C	K=10.
D	There is no such value of K.
190	For which value of K does the point $s = 0 + j\sqrt{15}$ lie on the root-locus of a system with the open loop transfer function $G(s) = \dfrac{K(s+5)}{s(s+1)(s+3)}$?
A	K=12.

B	K<20.
C	K=16.
D	There is no such value of K.
191	Is the root-locus method useful in designing a lead-lag controller for a system with desirable closed-loop transient response?
A	Yes, only if all the roots of characteristic equation are real.
B	Yes, only if all the roots of characteristic equation are complex.
C	Root-locus method is not suitable for this type of problems.
D	Root-locus method is suitable for designing a lead and/or lag controller to address this problem.
192	Controller, $G_c(s)$ needs to be chosen in such a way that the closed loop system exhibits stable behavior with $G_p(s) = \dfrac{K}{s(s-2)}$. Which type of controller $G_c(s)$ may be a good choice when the system does not have noisy signals at all?
A	A lag-network is best suited and hence $G_c(s)$ must be an implementation of lag network.
B	A lead-network is best suited hence $G_c(s)$ must be a lead-network.
C	A Proportional+Derivative controller of the form $G_c(s) = (K_p + K_d\, s)$ may solve the problem.
D	It is not possible to stabilize the system since $G_p(s)$ has a pole in the right hand side of the imaginary axis.
193	Does RL-method address the problem of indicating relative stability?
A	Just like Routh's criterion, RL-method addresses only absolute stability and does not indicate any method of avoiding instability.
B	RL-method is not a graphical method and so its use is minimal.
C	RL-method does indicate relative stability up to an extent in some situations.
D	The question is meaningless.
194	Is it advisable to choose value of K in a system with open loop transfer function of the form $G(s) = \dfrac{K(s+z_1)(s+z_2)\cdots(s+z_m)}{(s+p_1)(s+p_2)\cdots(s+p_n)}$ and H(s)=1 such that the Root Locus has a saddle point on the negative real axis?

A	It does not matter as long as this value of K results in pushing the roots of characteristic equation into the left half of s-plane.
B	As long as the settling time of the closed loop system is very small, it doesn't matter whether there is saddle point or not.
C	It is desirable that the Root Locus plot has a saddle point since the response will exhibit characteristics similar to that of a second order critically damped system.
D	It is not advisable to choose values of K near to that at saddle point since the sensitivity is very high at saddle point.
195	Which of the following statements about Root Locus plot is very appropriate?
A	Root Locus plot is always symmetric about the real-axis.
B	Extent of Root Locus on the real axis is independent of the poles not lying on the real axis.
C	A break-away point exists on the real axis when Root Locus exists between two adjacent poles on real axis.
D	All the above statements are true.
195	A closed-loop system has open loop transfer function $G(s) = \dfrac{K}{s^2(s+16)}$. Which of the following sketches shows the existence of roots of characteristic equation on real axis?
A	
B	

| C | 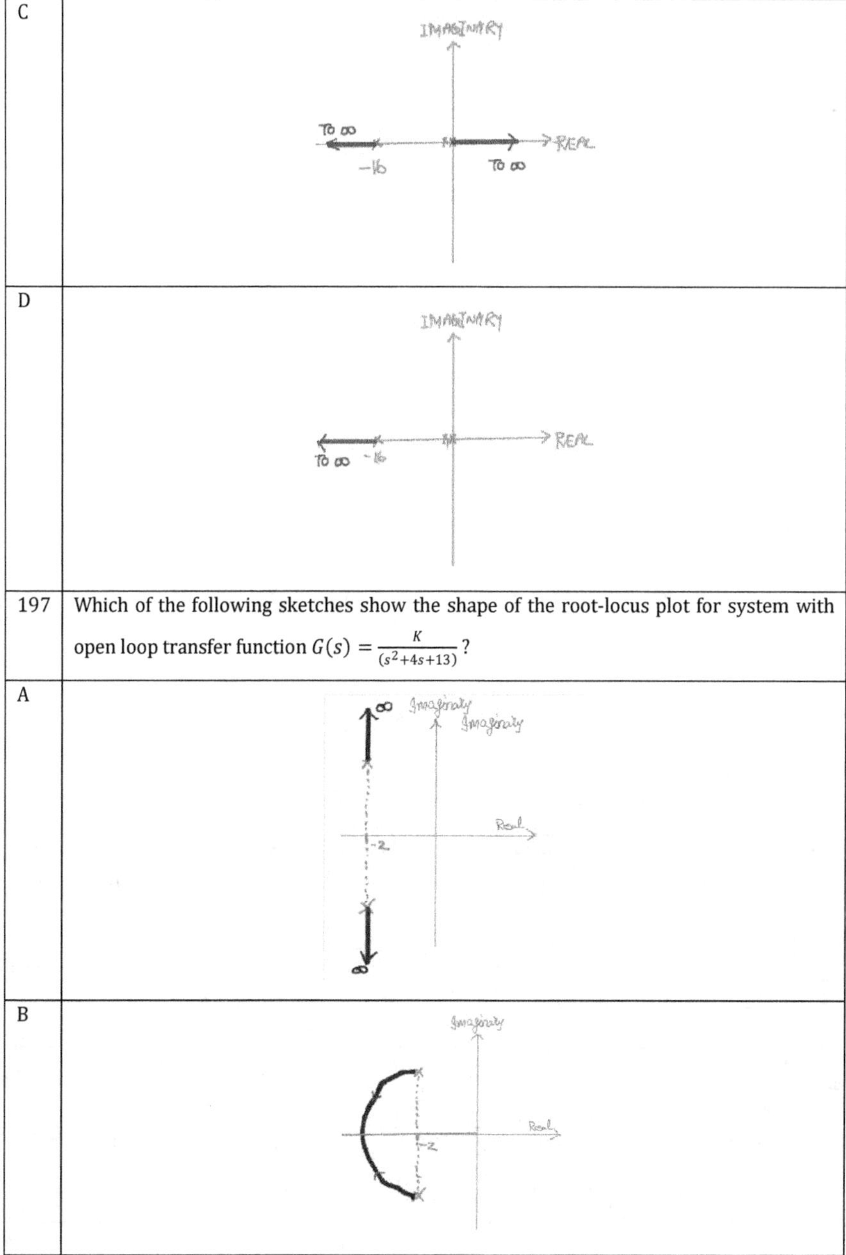 |

| D | |

| 197 | Which of the following sketches show the shape of the root-locus plot for system with open loop transfer function $G(s) = \dfrac{K}{(s^2+4s+13)}$? |

| A | |

| B | |

C	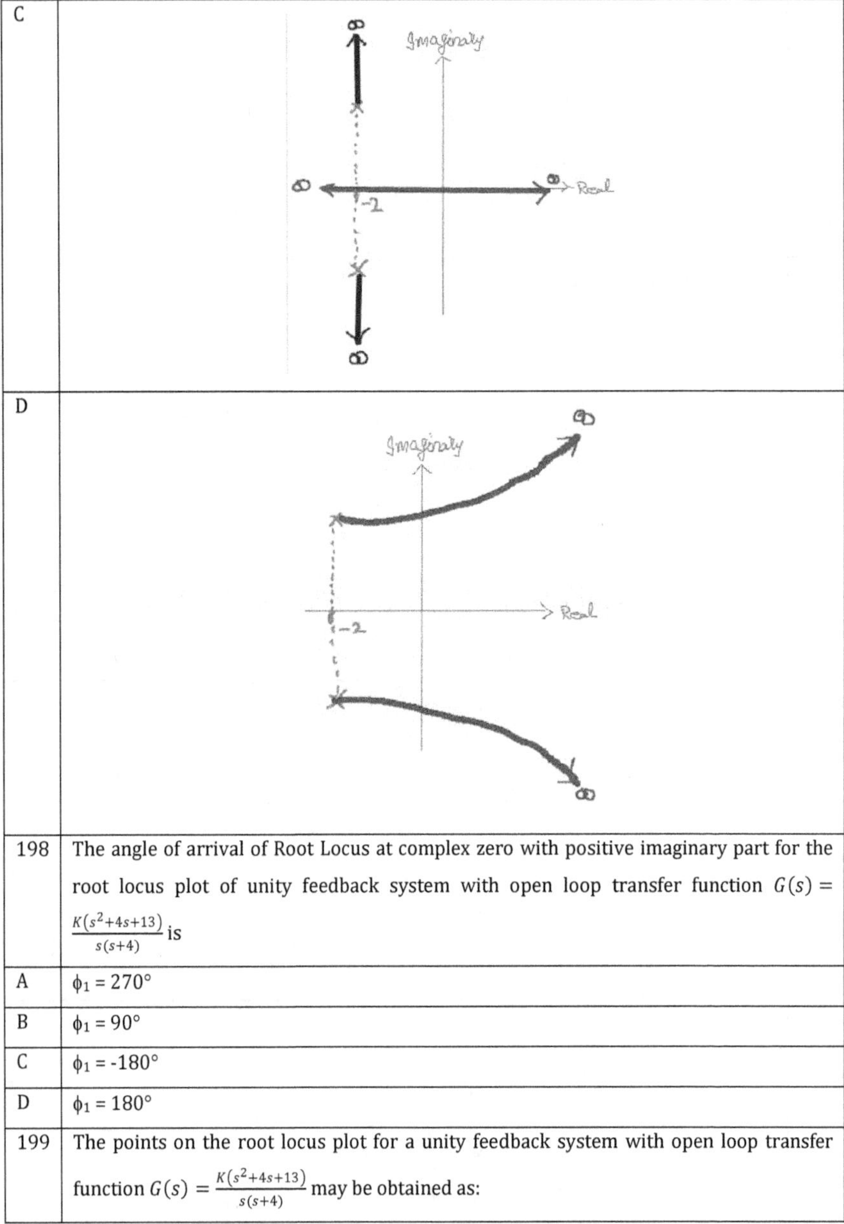
D	

198	The angle of arrival of Root Locus at complex zero with positive imaginary part for the root locus plot of unity feedback system with open loop transfer function $G(s) = \dfrac{K(s^2+4s+13)}{s(s+4)}$ is
A	$\phi_1 = 270°$
B	$\phi_1 = 90°$
C	$\phi_1 = -180°$
D	$\phi_1 = 180°$
199	The points on the root locus plot for a unity feedback system with open loop transfer function $G(s) = \dfrac{K(s^2+4s+13)}{s(s+4)}$ may be obtained as:

A	$$s_{1,2} = 2 \pm \sqrt{\dfrac{4 + 9K}{1 + K}}$$
B	$$s_{1,2} = 2 \pm \sqrt{\dfrac{4 - 9K}{1 + K}}$$
C	$$s_{1,2} = 2 \pm \sqrt{\dfrac{4 - 9K}{1 - K}}$$
D	$$s_{1,2} = 2 \pm \sqrt{\dfrac{4 + 9K}{1 - K}}$$
200	When $K = \dfrac{4}{9}$, the factored form of closed loop characteristic equation for the system with open loop transfer function $G(s) = \dfrac{K(s^2+4s+13)}{s(s+4)}$ is
A	$(s+2)^2 = 0.$
B	$(s^2 + 4s + 13) = 0.$
C	$(s^2 + 4s + 16) = 0.$
D	None of the above.
201	When $K = \dfrac{9}{4}$, the factored form of closed loop characteristic equation for the system with open loop transfer function $G(s) = \dfrac{Ks(s+2)}{(s^2+4s+13)}$ is
A	$(s+2)^2 = 0.$
B	$(s^2 + 4s + 13) = 0.$
C	$(s^2 + 4s + 16) = 0.$
D	None of the above.
202	What can be said about the root loci of systesms with open loop transfer functions $G_1(s) = \dfrac{Ks(s+4)}{(s^2+4s+13)}$ and $G_1(s) = \dfrac{K(s^2+4s+13)}{s(s+4)}$?
A	Both root loci are identical in all respects since poles of one transfer function are the zeros of the other transfer function.
B	Both root loci will look similar except for the locations of poles and zeros.
C	Root loci of both the transfer functions share the same saddle point. Other than this, everything else is different.
D	Nothing can be said about the similarity of root loci of the transfer functions.
203	Can saddle point of root locus of a system be a complex number?
A	When K is complex quantity, saddle point can be a complex number.

B	When angle of departure is complex, saddle point can be a complex number.
C	It is not possible to have a complex saddle point.
D	In some situations, saddle point can be a complex number.
204	A closed loop system is shown in figure below. What is the range of values of K for stability? 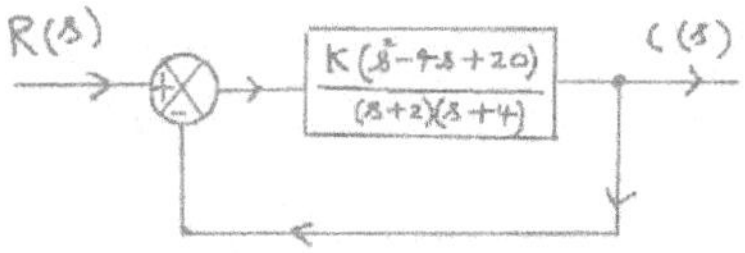 $R(s) \to \boxed{\dfrac{K(s^2-9s+20)}{(s+2)(s+4)}} \to C(s)$ (with unity feedback)
A	Since the open loop transfer function has non-minimum phase zero, all values of K will yield unstable system.
B	K needs to be very, very high for stability of the system.
C	K needs to be shown to be between 0 and 15 for stability.
D	All the above statements are incorrect.
205	For what value of K, the system shown in Figure below exhibits sustained oscillations? 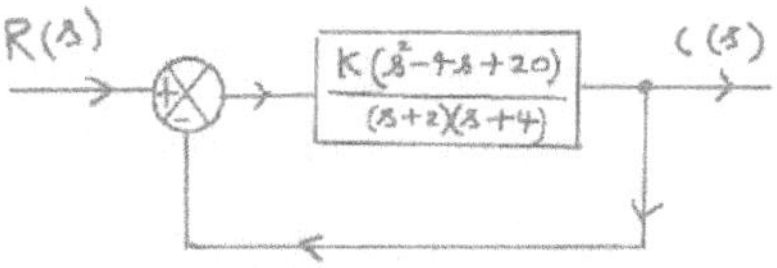 $R(s) \to \boxed{\dfrac{K(s^2-9s+20)}{(s+2)(s+4)}} \to C(s)$ (with unity feedback)
A	$K = \dfrac{3}{2}$
B	$K = \dfrac{2}{3}$
C	$K = 0$
D	Sustained oscillations are NOT possible in the system.
206	What is the frequency of sustained oscillations in the following system when such oscillations occur? 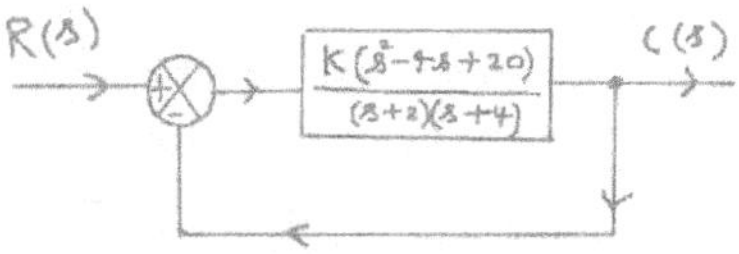 $R(s) \to \boxed{\dfrac{K(s^2-9s+20)}{(s+2)(s+4)}} \to C(s)$ (with unity feedback)
A	3.9 cycles/sec
B	3.9 rad/sec

C	9.3 rad/sec
D	Sustained oscillations are not possible in the system.
207	A dynamic system is described by the equation $\dot{x} = f(t,x)$ where t is the time elapsed and x_0 is the initial condition for the system. A point y is defined by the condition $f(t,y) = 0 \ \forall \ t \geq t_1$. Then the point y is termed as:
A	Equilibrium point of the system at time t = t_1.
B	Stability point of the system at time t = t_1.
C	Asymptotic point of the system at time t = t_1.
D	There is no name for the point y.
208	An equilibrium point x_e of the dynamic system described by the equation $\dot{x} = f(t,x), x(t_0) = x_0$ satisfies the following mathematical statement: $$\exists \, r > 0 \ni \nexists \, x \in B(x_e, r) \ni \{(f(t,x) = 0, t \geq t_1) \wedge (x \neq x_e)\}$$ where $B(x_e, r) = \{x \ni |x - x_e| \leq r\}$. Then the equilibrium point of the dynamic system is termed as:
A	Exponentially stable equilibrium point of the system.
B	Asymptotically stable equilibrium point of the system.
C	Stable isolated equilibrium point of the system.
D	Isolated equilibrium point of the system.
209	Pick the choice that best fills the blank in the statement: A dynamic system is described by $\dot{x} = f(t,x)$ with equilibrium point x=0. The equilibrium point x = 0 is stable if __________ where g(t,x) is solution of the equation $\dot{x} = f(t,x)$.
A	$\forall \varepsilon, \exists \delta(\varepsilon) \ni |x_0| < \delta(\varepsilon) \Rightarrow \{|g(t,x)| < \varepsilon \ \forall t \geq t_0\}$
B	$\forall \varepsilon, \exists \delta(\varepsilon) \ni |x_0| > \delta(\varepsilon) \Rightarrow \{|g(t,x)| < \varepsilon \ \forall t \geq t_1\}$
C	δ and ε are functions of each other.
D	δ and ε are both bounded.
210	Equilibrium for the equation $\dot{x} = ax$ can be specified as:
A	x = a is an equilibrium for the system.
B	x = 0 is one of the many equilibria.
C	x = a is one of the many equilibria.
D	x = 0 is the only equilibrium of the system.
211	The solution of the equation $\dot{x} = ax$ for the initial condition x = 2 at t = 0 is:
A	$x = 2e^{at}$.
B	$\log_e x = at + c$ where $c = \log_e 2$.

C	$$\log_e \left[\frac{x}{2}\right] = at$$
D	All the above represent valid solutions of the equation.
212	Let $\emptyset(t, t_0, x_0)$ be the solution of the dynamic system $\dot{x} = -2x$ starting from the initial condition x = 3 at t = 0. Then $\emptyset(t, t_0, x_0)$ may be given by:
A	$-2e^{3t}$
B	$2e^{-3t}$
C	$-3e^{-2t}$
D	$3e^{-2t}$
213	For the differential equation $\dot{x} = ax$ with the initial condition x = 2 at t = 0, the eqiulibrium point x = 0 is:
A	Stable $\forall$t.
B	Uniformly stable $\forall$t.
C	Exponentially stable in the large $\forall$a<0.
D	Uniformly asymptotically stable $\forall$a.
214	Consider the following assertions:
	(i) Lyapunov stability theory can be aplied only to nonlinear systems and does not apply to linear systems.
	(ii) Lyapunov stability theory is applicable only to linear systems.
	(iii) Lyapunov stability theory is applicable to both linear and nonlinear systems.
	(iv) Bounded-input-bounded-output stability theory applied to linear systems and Lyapunov stability theory applied to linear systems lead one to the same conclusion in a way.
	Now pick the choice that best applies regarding the assertions.
A	Assertions (i) and (ii) are correct and (iii) and (iv) are incorrect.
B	Assertions (i) and (iii) are correct and (ii) and (iv) are incorrect.
C	Assertions (i) and (ii) are incorrect and (iii) and (iv) are correct.
D	All the assertions are incorrect.
215	Consider the following assertions:
	(i) Adding a zero to OLTF has the effect of attracting the branches of root-loci towards it, in general.
	(ii) Adding a pole to OLTF has the effect of repelling the branches of root-loci away from it, in general.
	(iii) Adding poles and zeros to OLTF alters the root loci but in general root locus

	plot's shape does not change.
	(iv) The effects of addition poles and zeros to OLTF affect only the angle criterion.
	Now pick the choice that best applies.
A	Only (i) is correct.
B	Only (ii) is correct
C	(i) and (ii) are correct more or less.
D	Only (iii) and (iv) are correct.
216	Does the point s = -2 + 3j lies on the root locus plot for $G(s) = \dfrac{K}{s(s^2+8s+21)}$?
A	We can choose K suitably to ensure that s = -2 + 3j lies on the root locus.
B	We can choose K suitable to ensure that s = -2 + 3j lies on the root locus but we will have sustained oscillations.
C	There will be several values of K for which s = -2 + 3j lies on the root locus.
D	There is no value of K for which the point s = -2 + 3j lies on the root locus.
217	Pick the statement which is correct regarding root locus on real axis.
A	A point on the real axis lies on root locus if the sum of number of poles and zeros to the right of the point is an odd number.
B	A point on the real axis lies on root locus if the sum of number of poles and zeros to the right of the point is an even number.
C	A point on the real axis lies on root locus if the sum of poles and zeros to the right of the point is an odd number.
D	A point on the real axis lies on root locus if the sum of poles and zeros to the right of the point is an even number.
218	Which among the following is a unique model of a system?
A	Transfer function
B	State variable
C	Block diagram
D	Signal flow graphs
219	Which among the following is a disadvantage of modern control theory?
A	Implementation of optimal design
B	Transfer function can also be defined for different initial conditions
C	Analysis of all systems take place
D	Necessity of computational work

220	Which property in control engineering implies an ability to measure the state by taking measurements at output?
A	Controllability
B	Observability
C	Differentiability
D	Adaptability
221	Which of the following is valid root locus diagram?
A	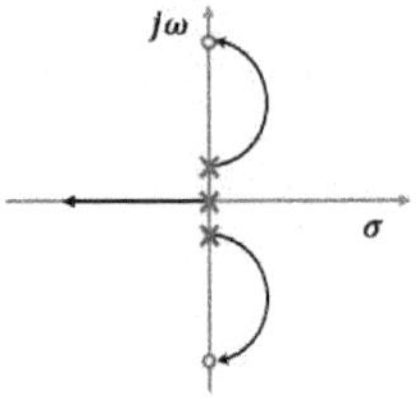
B	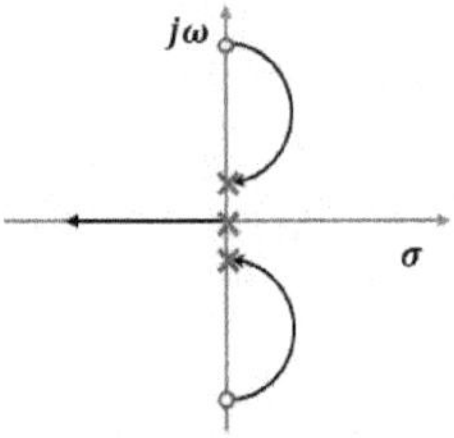
C	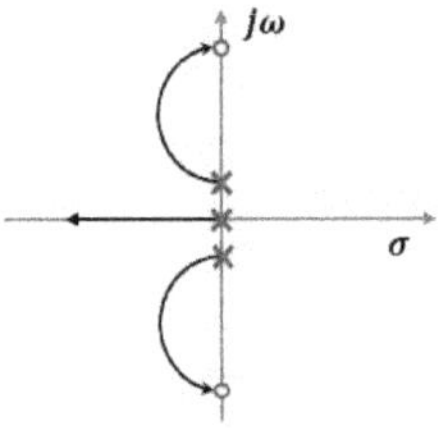
D	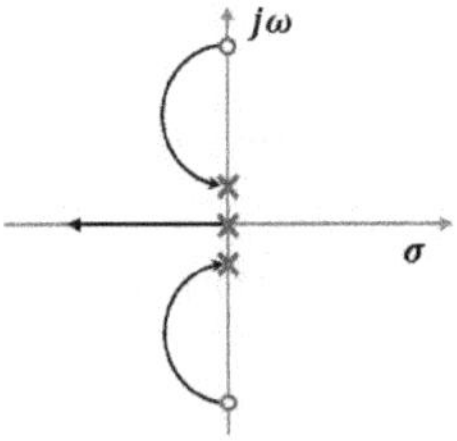

222	Consider the feedback system shown below. For this system, the root locus plot is::

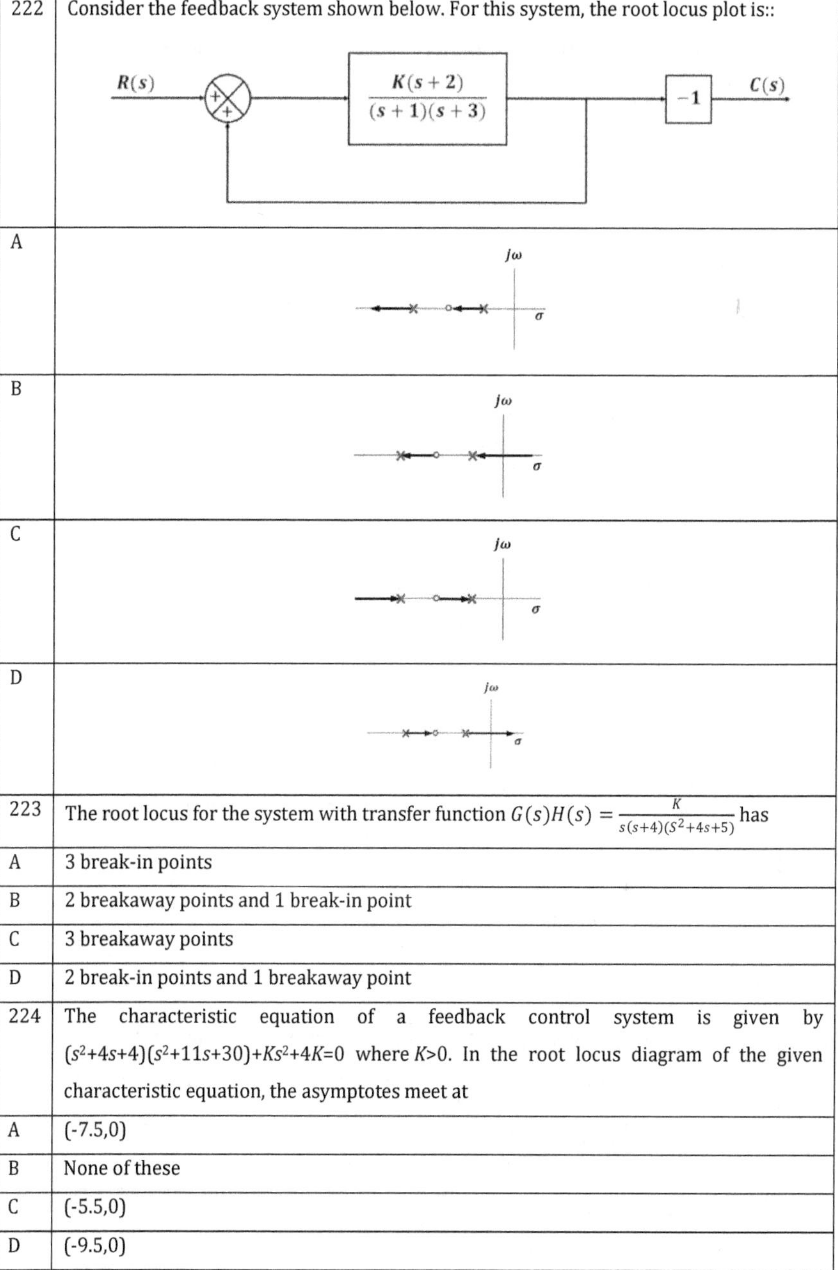

A	
B	
C	
D	

223	The root locus for the system with transfer function $G(s)H(s) = \dfrac{K}{s(s+4)(s^2+4s+5)}$ has
A	3 break-in points
B	2 breakaway points and 1 break-in point
C	3 breakaway points
D	2 break-in points and 1 breakaway point
224	The characteristic equation of a feedback control system is given by $(s^2+4s+4)(s^2+11s+30)+Ks^2+4K=0$ where $K>0$. In the root locus diagram of the given characteristic equation, the asymptotes meet at
A	(-7.5,0)
B	None of these
C	(-5.5,0)
D	(-9.5,0)

225	The forward path transfer function of a unity feedback system is $G(s) = \dfrac{K(s+3)}{s(s+1)(s+2)(s+4)}$. The angles of asymptotes in the root locus plot for this system are:
A	$\dfrac{\pi}{3}, \pi, \dfrac{5\pi}{3}$
B	$0, \dfrac{\pi}{2}, \dfrac{\pi}{3}$
C	$0, \dfrac{\pi}{3}, \dfrac{2\pi}{3}$
D	None of these
226	For the unity feedback system shown in Figure below consider two points $s_1 = -2 + j3$ and $s_2 = -2 + j\dfrac{1}{\sqrt{2}}$ 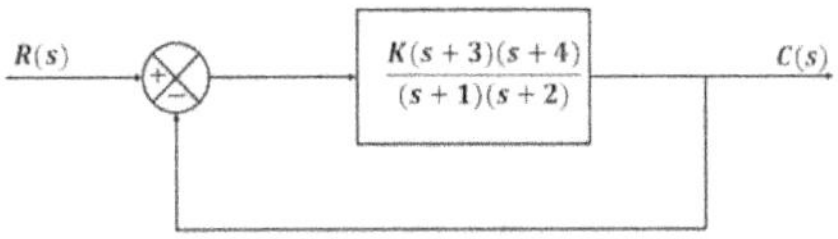
A	s_1 but not s_2
B	s_2 but not s_1
C	neither s_1 nor s_2
D	Both s_1 and s_2
226	Consider the unity feedback system with the open loop transfer function $G(s) = \dfrac{(s+a)}{s(s+b)}$. For this system, no part of the root locus lies in RHP. For this system, which condition must be satisfied so that the complex points of the root locus form a circle.
A	a > b is a necessary condition.
B	b > a is a sufficient condition.
C	a = b is necessary condition.
D	b > a is necessary condition.
227	Consider the unity feedback system with the open loop transfer function $G(s) = \dfrac{(s+a)}{s(s+b)}$. For this system, no part of the root locus lies in RHP. Further it is given that a > b. It is given that the complex points of the root locus form a circle. Then which of the following statement is true.
A	Center is (-a,0) and the radius is $\sqrt{b^2 - ab}$.
B	Center is (-b,0) and the radius is $\sqrt{b^2 - ab}$.

For question 226 (Figure):

$R(s)$ → ⊗ (+/−) → $\dfrac{K(s+3)(s+4)}{(s+1)(s+2)}$ → $C(s)$

C	Center is (-a,0) and the radius is $\sqrt{a^2 - ab}$.
D	Center is (-b,0) and the radius is $\sqrt{a^2 - ab}$.
228	Consider the unity feedback system with the open loop transfer function $G(s) = \frac{(s+6)}{s(s+4)}$. The values of K for which the system is underdamped satisfy the condition:
A	9.464>K>2.536
B	14.93>K>1.072
C	K>14.93 and K<1.072
D	K>9.464 and K<2.536
229	Consider the unity feedback system shown below. The root loci as α is varied are shown in: 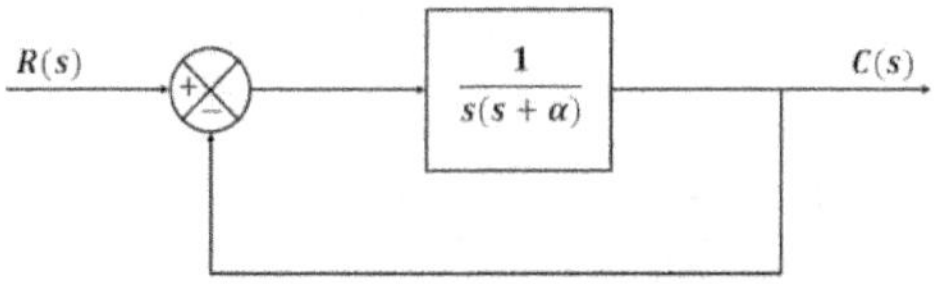

A	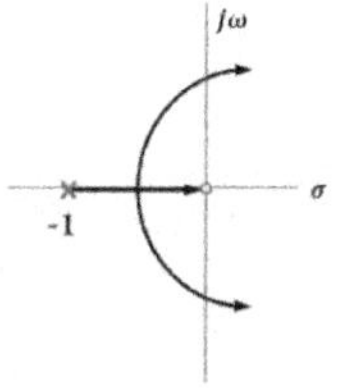
B	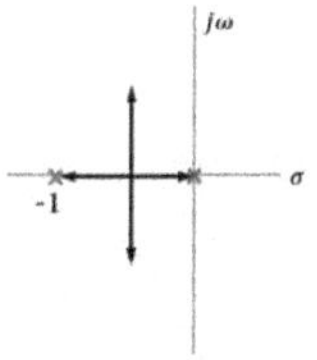
C	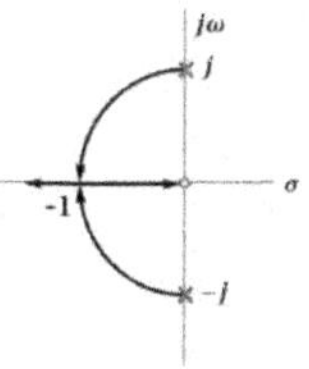

D	

230	The forward path transfer function of a unity feedback system is $G(s)H(s) = \dfrac{K(s+2)}{(s+3)(s^2+2s+2)}$. The angle of departure from the complex poles is
A	$\pm 108.4^o$
B	$\pm 203.4^o$
C	$\pm 136.4^o$
D	$\pm 213.4^o$
231	The characteristic equation of a closed loop system is $s(s+1)(s+3)+K(s+2)=0, K>0$. Which of the following is true?
A	It cannot have a breakaway point in the range $-1<Re(s)<0$
B	Two of its roots tend to infinity along the asymptotes $Re(s)=-1$
C	Its roots are always real
D	It may have complex roots in the right half plane
232	Table below shows the completed Routh's array for a closed loop system. Which of the following statements about the stability of the system is true?

1	3	3	1
2	4	2	
1	2	1	
2	2		
1	1		
2			
1			

A	The system is stable.
B	It is not possible to decide about the stability of the system.
C	The system is stable only in some situations
D	The system is always unstable.
233	The open-loop transfer functions with unity feedback are given below for different systems. The unstable system is
A	$\dfrac{2}{s+2}$

B	$\dfrac{2}{s^2(s+2)}$
C	$\dfrac{2(s+1)}{s(s+2)}$
D	$\dfrac{2}{s(s-2)}$

234	Consider a unity feedback system with feed forward transfer function $$G(s) = \frac{K(s+3)(s+5)}{(s-2)(s-4)}.$$ The range of K to ensure stability is
A	$K < -1 \text{ or } K > \dfrac{3}{4}$
B	$-1 < K < \dfrac{3}{4}$
C	$K > \dfrac{6}{8}$
D	$K < -1$

235	The closed loop transfer function of a system is $T(s) = \frac{s^3+4s^2+8s+16}{s^5+3s^4+5s^2+s+3}$. The number of poles in the right-half plane and left-half plane are
A	3,2
B	1,4
C	2,3
D	4,1

236	The closed loop transfer function of the system shown in the following block diagram is

A	$\dfrac{K}{s^3 + 5s^2 + 7s + K}$
B	$\dfrac{K}{s^3 + 5s^2 + 6s + K}$
C	$\dfrac{K}{s(s^2 + 5s + 7)}$
D	$\dfrac{K}{s(s^2 + 5s + 6)}$

237	For the system shown in block diagram below, the range of gain K for stability is:

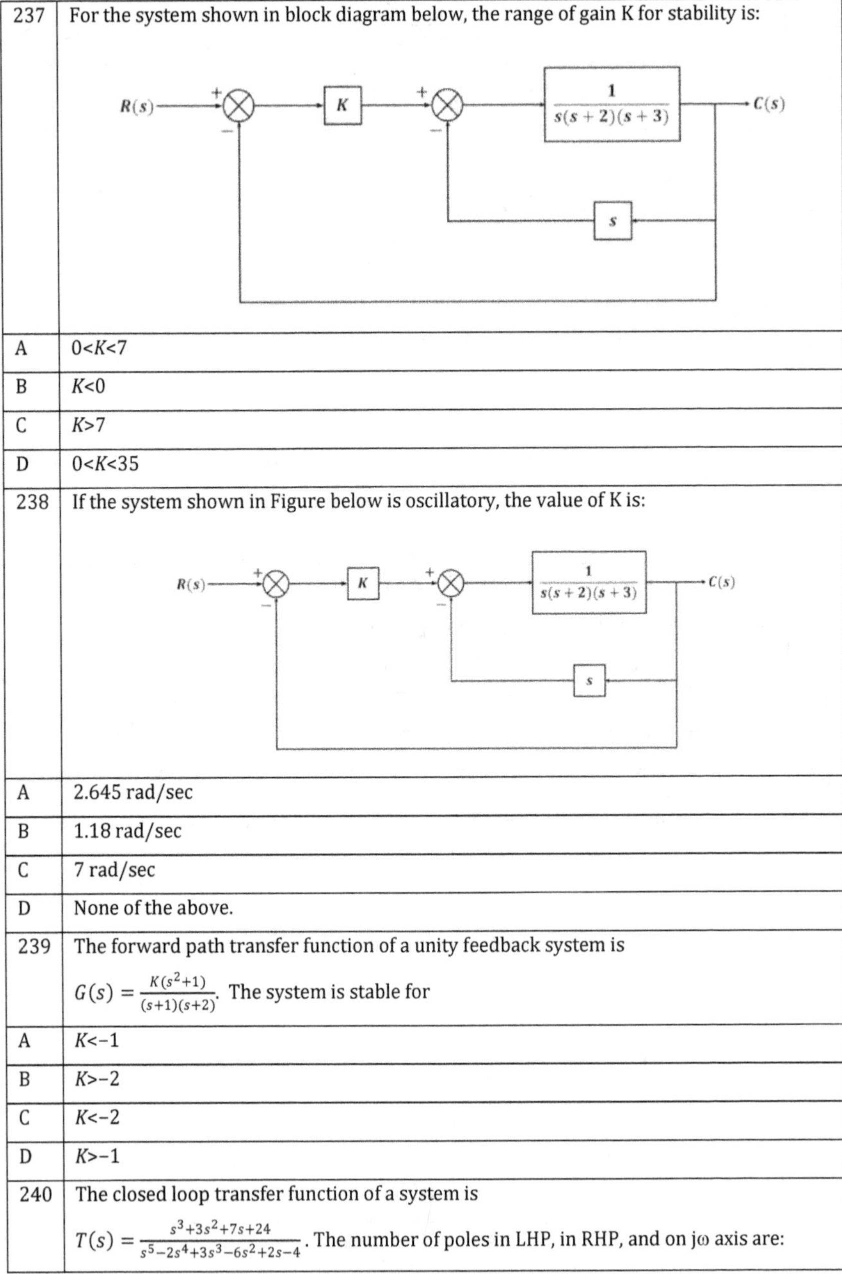

A	0<K<7
B	K<0
C	K>7
D	0<K<35

238	If the system shown in Figure below is oscillatory, the value of K is:

A	2.645 rad/sec
B	1.18 rad/sec
C	7 rad/sec
D	None of the above.

239	The forward path transfer function of a unity feedback system is $G(s) = \dfrac{K(s^2+1)}{(s+1)(s+2)}$. The system is stable for

A	$K<-1$
B	$K>-2$
C	$K<-2$
D	$K>-1$

240	The closed loop transfer function of a system is $T(s) = \dfrac{s^3+3s^2+7s+24}{s^5-2s^4+3s^3-6s^2+2s-4}$. The number of poles in LHP, in RHP, and on jω axis are:

A	1,2,2
B	0,1,4
C	1,0,4
D	2,1,2
241	The system shown in figure below is

A	Stable for input u_1 but unstable for input u_2
B	Stable
C	Conditionally Stable
D	Unstable
242	Find the range of K for which a system with characteristic equation $s^4+Ks^3+2s^2+(K+1)s+10=0$ is stable.
A	$K>1$
B	Not stable for real values of K
C	$\dfrac{1}{3} > K > -\dfrac{1}{3}$
D	$1 > K > \dfrac{1}{3}$
243	The open loop transfer function of a unity feedback system is $G(s) = \dfrac{4}{s(s^2+qs+2p)}$. Find the values of q and p such that the system oscillates at a frequency of 4 rad/sec.
A	$p=0.5,q=4$
B	$p=4,q=1$
C	$p=8,q=0.25$
D	$p=1,q=2$

244	For the unity feedback system whose open loop transfer function is $$G(s) = \frac{K}{s(s+1)(s+2)(s+5)}$$ the actual location of the closed loop pole(s), determined by using Routh-Hurwitz criterion, when the system is marginally stable is(are)
A	The system cannot be marginally stable
B	$s=-3.5,-4.5,\pm j1.118$
C	$s=-3.5,-4.5$
D	$s=\pm j1.118$
245	An open loop pole-zero plot is shown below. The transfer function of the system with the above pole-zero plot is
A	$\dfrac{K(s^2 - 2s + 2)}{(s+3)(s+2)}$
B	$\dfrac{K(s^2 + 2s + 2)}{(s+3)(s+2)}$
C	$\dfrac{K(s+3)(s+2)}{(s^2 - 2s + 2)}$
D	$\dfrac{K(s+3)(s+2)}{(s^2 12s + 2)}$
246	For the system with pole-zero plot shown in figure below, the break point is
A	Break-away at $\sigma=-2.43$

B	Break-in at $\sigma=-2.43$
C	Break-away at $\sigma=-1.29$
D	break-in at $\sigma=-1.29$
247	The open loop transfer function $G(s)$ of a unity feedback control system is given as $G(s) = \dfrac{K\left(s+\frac{2}{3}\right)}{s^2(s+2)}$. From the root locus, it can be inferred that when K tends to positive infinity, ________(Choose the correct option to fill in the blank)
A	Three real roots are found on the right half of the s-plane
B	Three roots with nearly equal real parts exist on the left half of the s-plane
C	One real root is found on the right half of the s-plane
D	The root loci cross the $j\omega$ axis for a finite value of K; $K\neq0$
248	An open loop pole-zero plot is shown below. 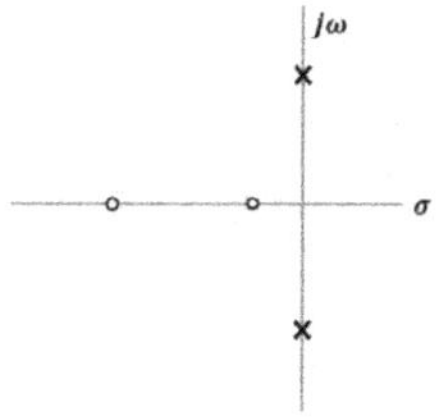 The general shape of the root locus plot is:
A	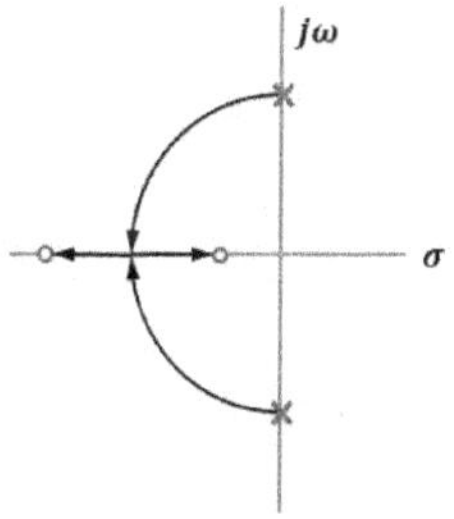
B	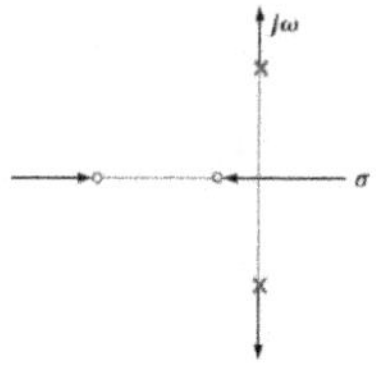

C	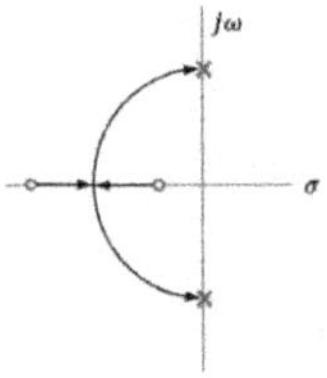
D	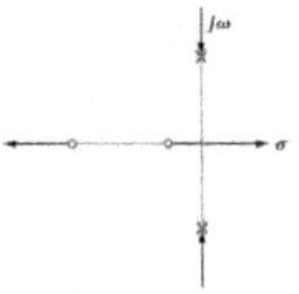

249	The forward-path transfer function of a unity feedback system is

$$G(s) = \frac{K(s+1)(s+2)}{(s+5)(s+6)}.$$

The break points in the root locus plot are

A	Break-in at -1.563 and break-away at -5.437
B	Break-in at -1.216 and break-away at -5.74
C	Break-in at -5.743 and break-away at -5.216
D	Break-in at -5.437 and break-away at -1.563

250	The open loop transfer function of a unity feedback system is given by $G(s) = \dfrac{2(s+\alpha)}{s(s+2)(s+10)}$. Consider the root locus of the system with α as the parameter. Intercepts of asymptotes at the real axis is
A	-8
B	-6
C	$-\dfrac{10}{3}$
D	-4

251	The loop transfer function of a third-order system is given as

$$G(s)H(s) = \frac{K}{(1+sT_1)(1+sT_2)(1+sT_3)}.$$

Which of the following statement is true?

A	The system is always stable for any $T_1, T_2, T_3, K \in [0,\infty)$.
B	The system is stable if $T_1, T_2, T_3, K \in [0,\infty)$ and $\left\| G\left(j\dfrac{1}{\sqrt{T_1 T_2 + T_2 T_3 + T_3 T_1}} \right) \right\| < 1$.

C	The system is stable if $T_1, T_2, T_3, K \in [0, \infty)$ and $\left| G\left(j\sqrt{\frac{T_1 + T_2 + T_3}{T_1 T_2 T_3}} \right) \right| < 1$.
D	The system is stable if $T_1, T_2, T_3, K \in [0, \infty)$ and both $\left| G\left(j\frac{1}{\sqrt{T_1 T_2 + T_2 T_3 + T_3 T_1}} \right) \right| < 1$, $\left| G\left(j\sqrt{\frac{T_1 + T_2 + T_3}{T_1 T_2 T_3}} \right) \right| < 1$ conditions are satisfied.

252	The asymptotic approximation of the log-magnitude versus frequency plot of a certain system is shown below

A	$$\frac{20(s + 5)}{s(s + 2)(s + 25)}$$
B	$$\frac{20(s + 5)}{s^2(s + 2)(s + 25)}$$
C	$$\frac{10s^2(s + 5)}{(s + 2)(s + 25)}$$
D	$$\frac{50(s + 5)}{s^2(s + 2)(s + 25)}$$

253	The asymptotic approximation of Bode magnitude plot of $H(j\omega) = \dfrac{10^4(1 + j\omega)}{(10 + j\omega)(100 + j\omega)^2}$ is

A	

B	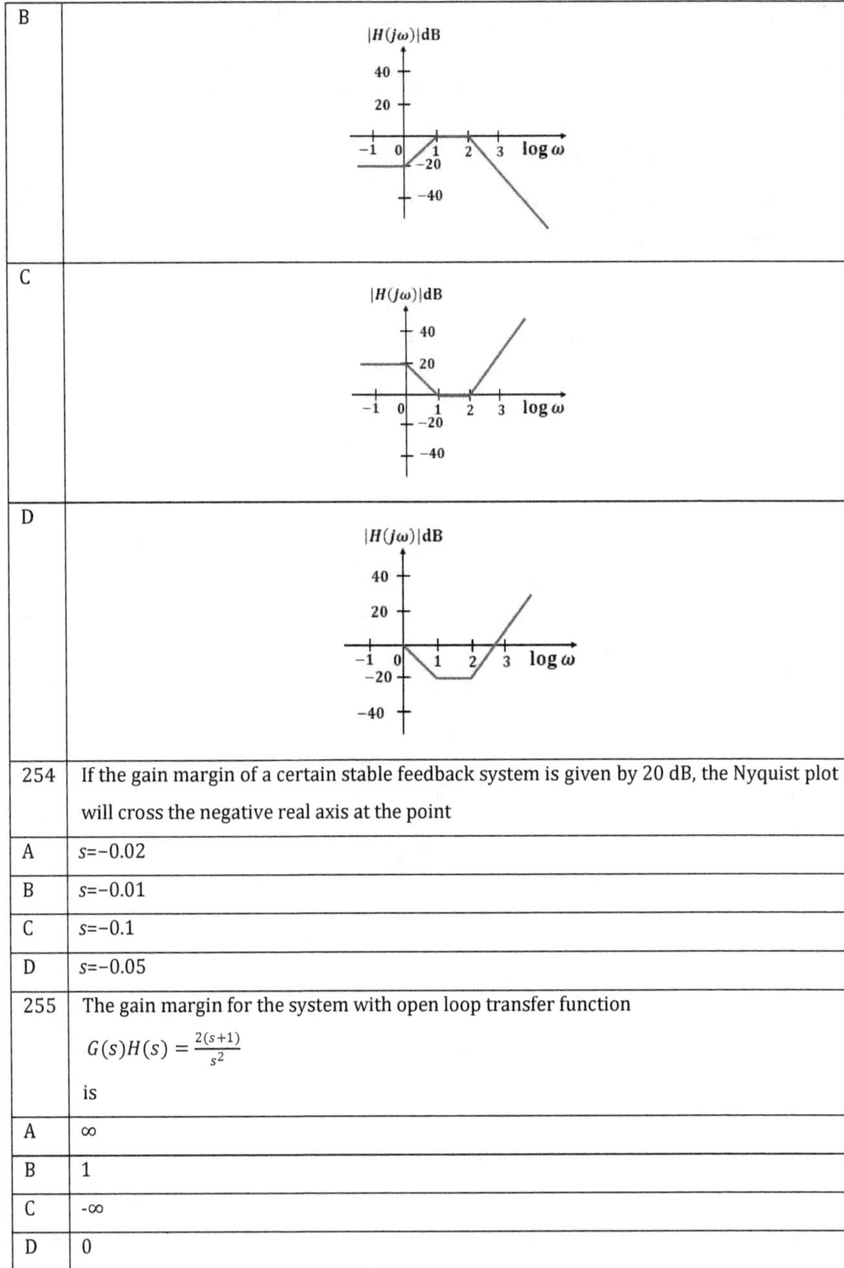
C	
D	

254	If the gain margin of a certain stable feedback system is given by 20 dB, the Nyquist plot will cross the negative real axis at the point
A	$s=-0.02$
B	$s=-0.01$
C	$s=-0.1$
D	$s=-0.05$
255	The gain margin for the system with open loop transfer function $$G(s)H(s) = \frac{2(s+1)}{s^2}$$ is
A	∞
B	1
C	$-\infty$
D	0

256	The open loop transfer function of a unity feedback system is given by $G(s) = \dfrac{3e^{-2s}}{s(s+2)}$. The gain crossover frequency is
A	$0.945 \ rad/s$
B	$0.485 \ rad/s$
C	$0.632 \ rad/s$
D	$1.26 \ rad/s$
257	The Nyquist plot of a feedback system is shown below.

Im(s)

−10.64

−40   −20   Re(s)

The open-loop transfer function is $G(s) = \dfrac{(4s+1)}{s^2(s+1)(2s+1)}$. The number of poles of the closed loop system in *RHP* is

A	2
B	1
C	0
D	4
258	The low frequency magnitude curve of the forward transfer function of a unity feedback system is given below

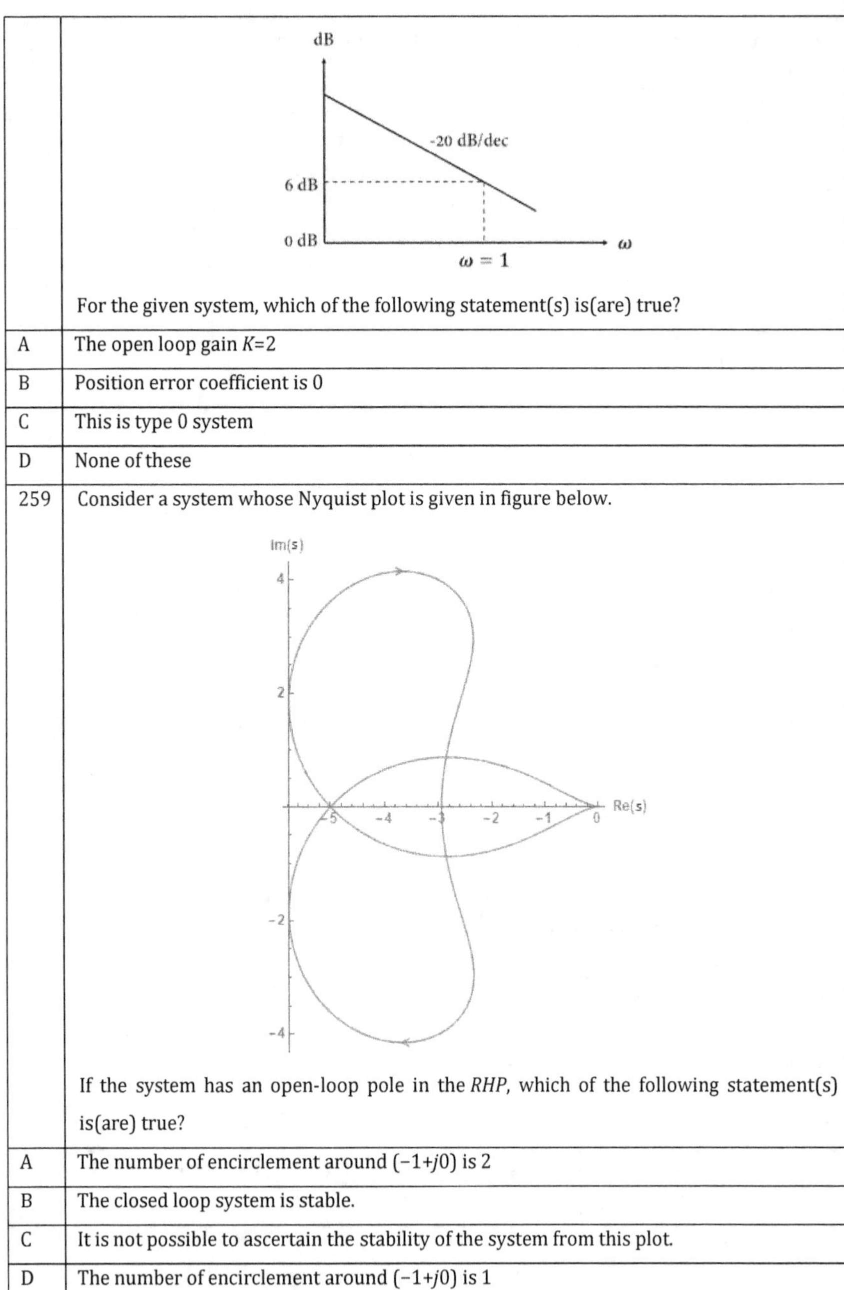

For the given system, which of the following statement(s) is(are) true?

A	The open loop gain $K=2$
B	Position error coefficient is 0
C	This is type 0 system
D	None of these

259	Consider a system whose Nyquist plot is given in figure below.

If the system has an open-loop pole in the *RHP*, which of the following statement(s) is(are) true?

A	The number of encirclement around $(-1+j0)$ is 2
B	The closed loop system is stable.
C	It is not possible to ascertain the stability of the system from this plot.
D	The number of encirclement around $(-1+j0)$ is 1

260	What is the *absolute value* of the real part of the dominant poles of the system with characteristic equation $s^3+14s^2+45s+50=0$?
A	2
B	3
C	4
D	5
261	Using integral control results in:
A	Improvement in stability of the system
B	Improvement in steady state error
C	Introduces damping in system
D	Reduces overshoot
262	Which of the following are false for the lag compensator?
A	Approximates a PD controller
B	Approximates a PI controller
C	Pole of the compensator is closer to the origin than the zero of the compensator
D	None of the above
263	A system with unity feedback has a open-loop transfer function of $G(s) = \dfrac{K(s+1)}{s(s-1)(s^2+4s+16)}$. When K=10, the value of G(s) when s = 1+j is given by
A	0
B	1+j
C	2+2j
D	$-\dfrac{10}{533}(44+j27)$
264	A system with unity feedback has a open-loop transfer function of $G(s) = \dfrac{K(s+1)}{s(s-1)(s^2+4s+16)}$. When K=10, the value of $|G(s)|$ when s = 1+j is given by
A	0
B	$\dfrac{10\sqrt{2665}}{533}$
C	1+j
D	2+2j
265	A system with unity feedback has a open-loop transfer function of $G(s) = \dfrac{K(s+1)}{s(s-1)(s^2+4s+16)}$. When K=10, the value of $\angle G(s)$ when s = 1+j is given by

A	$180 + \tan^{-1}\left(\dfrac{27}{44}\right)$
B	$180 - \tan^{-1}\left(\dfrac{27}{44}\right)$
C	$\tan^{-1}\left(\dfrac{27}{44}\right)$
D	$-\tan^{-1}\left(\dfrac{27}{44}\right)$
266	Consider a system with the transfer function $G(s) = \dfrac{2500}{s^2-1000}$. For this system, what kind of compensator will achieve a faster transient response?
A	Lead
B	Lag
C	Lead-Lag
D	None of the above
267	Consider a system with the transfer function $G(s) = \dfrac{2500}{s^2-1000}$. For this system, with a compensator of the form $\dfrac{K(s+z)}{(s+p)}$ with K, z, p > 0, what is the value of the velocity error constant K_v?
A	0
B	$\sqrt{1000}$
C	$\dfrac{25K}{zp}$
D	None of these
268	The forward path transfer function of a unity feedback system is $G(s) = \dfrac{K}{s(s+6.54)}$. When K=100, the resonant frequency is
A	8.86 rad/sec
B	18.86 rad/sec
C	188.86 rad/sec
D	1888.86 rad/sec
269	The forward path transfer function of a unity feedback system is $G(s) = \dfrac{K}{s(s+6.54)}$. When K=100, the resonant peak is
A	2
B	3
C	1.61
D	None of these

270	The forward path transfer function of a unity feedback system is $G(s) = \dfrac{K}{s(s+6.54)}$. When K=100, the bandwidth of the closed loop system is
A	14.3 rad/sec
B	123.7 rad/sec
C	0.1 rad/sec
D	5 rad/sec
271	Figure below shows the frequency response plot of a second order system. 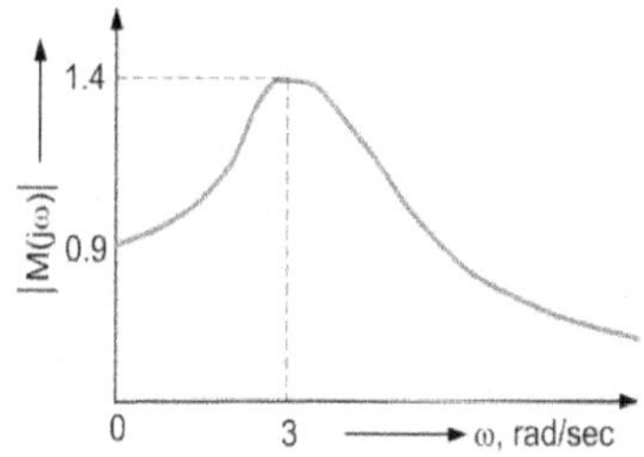 The approximate value of damping ratio of this system is
A	0.39 rad/sec
B	1.39 rad/sec
C	2.39 rad/sec
D	None of these
272	Figure below shows the frequency response plot of a second order system. 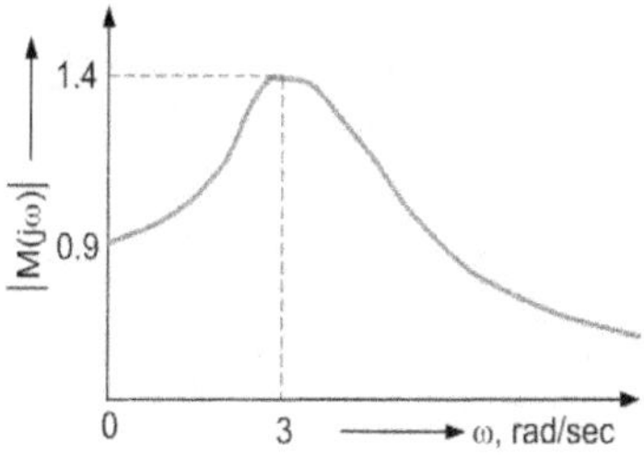 The approximate value of undamped natural frequency of the system is
A	6.59 rad/sec
B	4.59 rad/sec
C	5.59 rad/sec
D	3.59 rad/secc

273	Figure below shows the frequency response plot of a second order system.
	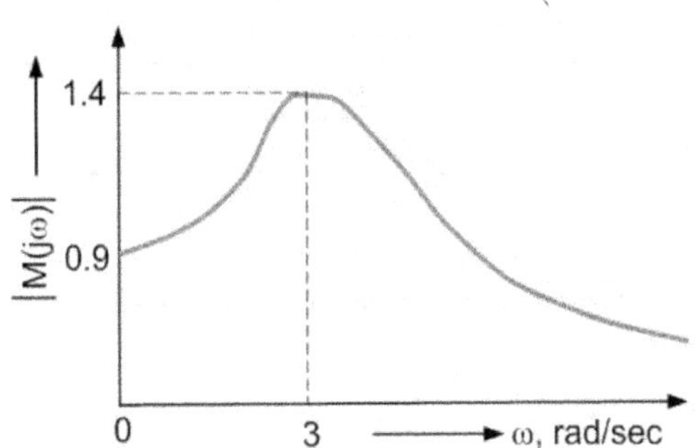
	The approximate value of %peak overshoot when this system is subjected to a unit step input is
A	26.7%
B	36.7%
C	46.7%
D	56.7%
274	Figure below shows the frequency response plot of a second order system.
	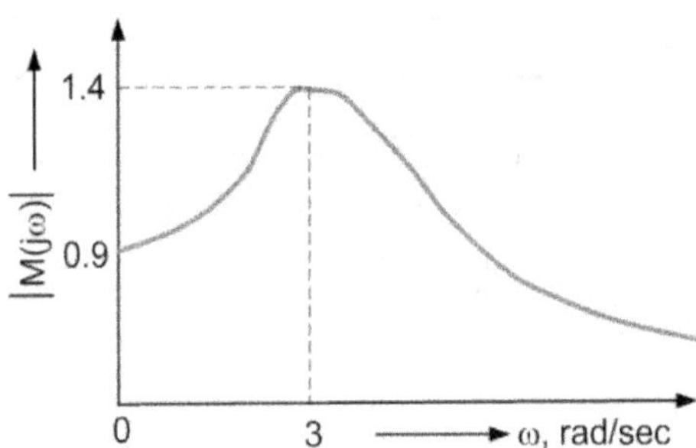
	The approximate value of 2% settling time of the system when this system is subjected to a unit step input is
A	2.88 sec
B	3.88 sec
C	4.88 sec
D	5.88 sec
275	Consider the asymptotic Bode plot in the figure below.

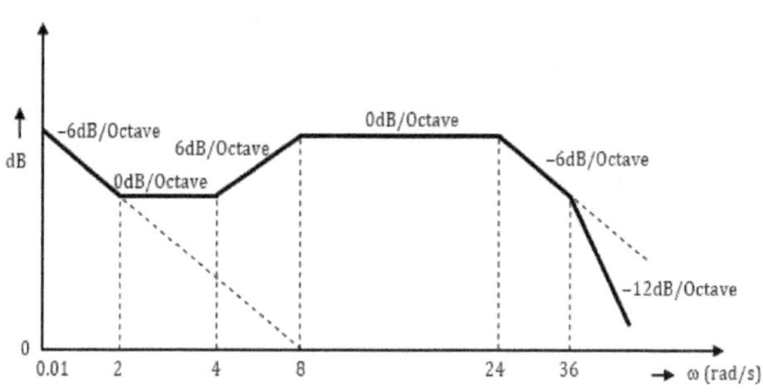

The transfer function is given by:

$$G(s) = \frac{K(1 + as)(1 + 0.25s)}{s(1 + \frac{s}{8})(1 + bs)(1 + \frac{s}{36})}$$

What is the value of a in the transfer function?

A	$\dfrac{1}{2}$
B	$\dfrac{1}{24}$
C	$\dfrac{1}{8}$
D	None of these

276	Consider the asymptotic Bode plot in the figure below.

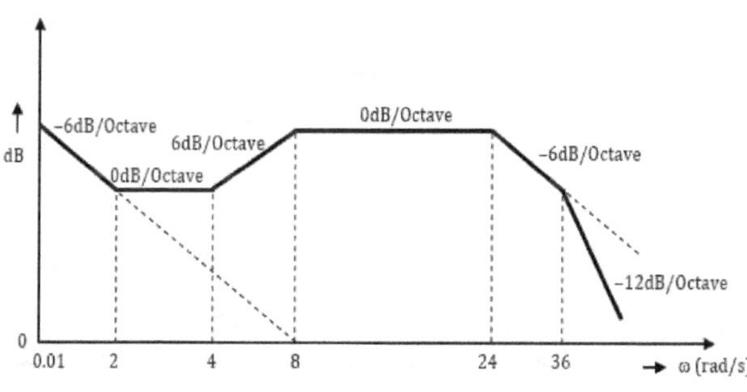

The transfer function is given by:

$$G(s) = \frac{K(1 + as)(1 + 0.25s)}{s(1 + \frac{s}{8})(1 + bs)(1 + \frac{s}{36})}$$

What is the value of b in the transfer function?

A	$\dfrac{1}{24}$
B	$\dfrac{1}{12}$
C	$\dfrac{1}{6}$
D	None of the above
277	Consider the asymptotic Bode plot in the figure below. 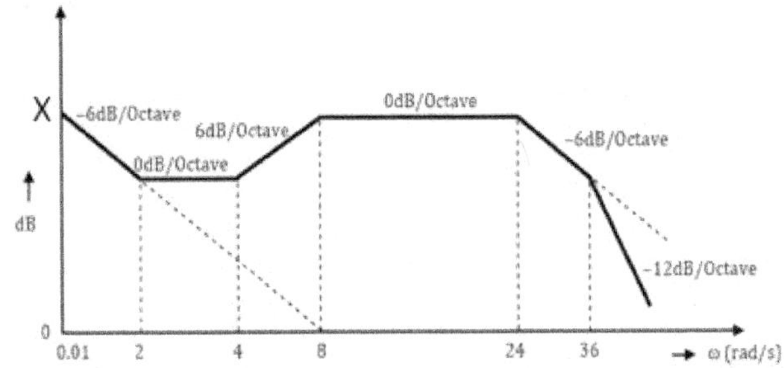 The transfer function is given by: $$G(s) = \dfrac{K(1 + as)(1 + 0.25s)}{s(1 + \frac{s}{8})(1 + bs)(1 + \frac{s}{36})}$$ What is the value of X in the plot?
A	27.41
B	17.41
C	37.41
D	47.41
278	Which among the following is a unique model of a system?
A	Transfer function
B	State variable
C	Block diagram
D	Signal flow graphs
279	Which among the following is a disadvantage of modern control theory?
A	Implementation of optimal design
B	Transfer function can also be defined for different initial conditions
C	Analysis of all systems take place
D	Necessity of computational work

280	Which property in control engineering implies an ability to measure the state by taking measurements at output?
A	Controllability
B	Observability
C	Differentiability
D	Adaptability